D0546091

Head Start to
A-Level Maths

Looking to start your A-Level Maths studies early? Well, let's get this show on the road with this unbeatable CGP Head Start book (including free Online Edition)!

Step 1: Recap key GCSE concepts (in case you've already forgotten!).
Step 2: Introduce new A-Level ideas with **clear study notes** and **practice questions**.
Step 3: Strut into A-Level Maths with confidence!

How to access your free Online Edition

This book includes a free Online Edition to read on your PC, Mac or tablet.
You'll just need to go to **cgpbooks.co.uk/extras** and enter this code:

4465 3147 1746 0911

By the way, this code only works for one person. If somebody else has used this book before you, they might have already claimed the Online Edition.

CGP — still the best! ☺

Our sole aim here at CGP is to produce the highest quality books —
carefully written, immaculately presented and dangerously close to being funny.

Then we work our socks off to get them out to you
— at the cheapest possible prices.

Contents

Section 5 — Graphs

Section 6 — Trigonometry and Vectors

Section 7 — Statistics and Probability

Throughout the book, we've marked up content that you're not likely to have come across at GCSE with stamps like this one. This is the material you'll be learning more about as part of A-level Maths.

Published by CGP

Typesetters: Callum Lamb, Kirsty Sweetman

Editors: Rachel Craig-McFeely, Sarah Pattison

ISBN: 978 1 78294 792 9

With thanks to Emma Clayton for the proofreading.
With thanks to Emily Smith for the copyright research.

Clipart from Corel®
Printed by Elanders Ltd, Newcastle upon Tyne.

Based on the classic CGP style created by Richard Parsons.

Text, design, layout and original illustrations © Coordination Group Publications Ltd. (CGP) 2021
All rights reserved.

Photocopying more than one section of this book is not permitted, even if you have a CLA licence.
Extra copies are available from CGP with next day delivery • 0800 1712 712 • www.cgpbooks.co.uk

Diagnostic Test

This first exercise will help you find out which areas of maths you need to work on before you start your A-Level Maths course. Do it before you work through the book — if you struggle with any of the questions, go straight to the relevant pages to brush up on your skills.

Once you've done all that, work through the rest of the book. You'll be able to recap and practise some useful GCSE topics and see how they'll lead into your A-Level work.

Types of Number and Fractions

These topics are covered in Section 1 — p.6-7.

1) Which of the following are integers?

4   −3.5   0.3 $\frac{4}{5}$ 8.99   −10   205   0

2) Which of the following values are rational, and which are irrational?

$5.\dot{9}$ π $\sqrt{7}$ $\frac{1}{5}$ −6 $\sqrt{4}$ 13.978   2.1

3) Evaluate the following without using a calculator, giving your answers in their lowest terms. Give any answers larger than 1 as improper fractions.

a) $\frac{2}{9} \times \frac{3}{5}$ b) $\frac{1}{6} \div \frac{2}{3}$ c) $\frac{1}{12} + \frac{5}{6}$ d) $\frac{8}{5} - \frac{1}{7}$

Indices, Multiplying Out Brackets and Factorising

These topics are covered on p.8-11.

4) Simplify the following:

a) $x^7 \times x^2$ b) $10y^3 \div 5y$ c) m^0 d) $(2n^2)^5$

5) Write 5^{-2} as a fraction.

6) Evaluate the following without using a calculator:

a) $\left(\frac{3}{4}\right)^2$ b) $16^{\frac{1}{2}}$ c) $8^{\frac{2}{3}}$ d) $36^{-\frac{1}{2}}$

7) Multiply out the brackets and simplify your answers where possible.

a) $(x + 4)(x - 6)$ b) $(x + 5)^2$ c) $(2x - 1)(x + 3)$ d) $(x + 1)(x - 4)(x + 5)$

8) Factorise the following:

a) $5x + 20$ b) $3a + 12ab$ c) $x^2 - 4$ d) $9x^2 - 36$ e) $x^2 - 5$

Surds

This topic is covered on p.12-13.

9) Simplify the following:

a) $\sqrt{3} \times \sqrt{2}$ b) $(\sqrt{5})^2$ c) $\frac{\sqrt{30}}{\sqrt{6}}$ d) $\sqrt{12} + 2\sqrt{3}$ e) $(1 + \sqrt{7})^2$

10) Rationalise the denominators of the following:

a) $\frac{3}{\sqrt{2}}$ b) $\frac{\sqrt{5}}{2\sqrt{2}}$ c) $\frac{2}{3 + \sqrt{6}}$ d) $\frac{\sqrt{2}}{1 - \sqrt{5}}$

TEST YOURSELF # Diagnostic Test

Solving Equations and Rearranging Formulas

You'll find these on p.14-15.

11) Solve the following:

    a) $5x - 2 = 8$ b) $3(x - 6) = 2(x - 4)$ c) $\dfrac{x+2}{3} + \dfrac{2x}{5} = x + 2$ d) $2x(x + 1) = 2x + 18$

12) Make x the subject of the following formulas:

    a) $y = mx + c$ b) $y = \dfrac{3x+2}{5}$ c) $y = 2x^2z + 1$ d) $y = \dfrac{3x+1}{x-2}$

Quadratic Equations

Quadratics are covered in Section 3 — p.16-21.

13) Solve the following by factorising:

    a) $x^2 - 3x + 2 = 0$ b) $x^2 + 6x + 5 = 0$ c) $2x^2 - 3x - 5 = 0$ d) $3x^2 - 13x = -12$

14) Solve the following using the quadratic formula. Give your answers to two decimal places.

    a) $x^2 + 2x - 10 = 0$ b) $2x^2 - 5x - 1 = 0$

The formula is: $\dfrac{-b \pm \sqrt{b^2 - 4ac}}{2a}$.

15) Solve the following by completing the square. Give your answers as surds.

    a) $x^2 - 4x - 2 = 0$ b) $2x^2 + 4x - 7 = 0$

16) a) Complete the square for $x^2 + 6x + 8$.

    b) Hence sketch the graph of $y = x^2 + 6x + 8$, labelling the turning point and intercepts with the x-axis.

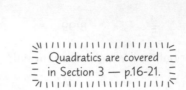

Helen thought that her diagnosis was ab-surd.

Algebraic Fractions, Inequalities and Simultaneous Equations

17) Simplify the following:

    a) $\dfrac{15a^3b^3}{5a^2b}$ b) $\dfrac{2x^2y}{(4xy)^2}$ c) $\dfrac{x^2 - 16}{x^2 - x - 20}$

These topics are on p.22-29.

18) Simplify the following:

    a) $\dfrac{9b^2}{a} \times \dfrac{2a^2}{3b}$ b) $\dfrac{2(x-1)^2}{15} \times \dfrac{10}{4x-4}$ c) $\dfrac{3x^2 - 21x}{x+2} \div \dfrac{x(x-7)}{9x+18}$ d) $\dfrac{3}{x+1} + \dfrac{2x-3}{x^2}$

19) Solve the following inequalities:

    a) $7x + 5 \le 2x$ b) $2(10 - x) > 4$ c) $2x^2 + 3 < 21$
    d) $4x^2 - 9 \ge 7$ e) $x^2 - 4x + 10 \ge 2x + 5$

20) Draw a set of axes with the x-axis from -2 to 3 and the y-axis from 0 to 6. Show on these axes the region that satisfies the following inequalities:

$$y > 3x - 1, \qquad y < x + 3 \qquad \text{and} \qquad y \ge \frac{x}{5} + 2$$

21) Solve the following simultaneous equations:

    a) $2x + y = 2$ b) $3x - 2y = 1$ c) $y = x^2 + 3$ d) $3y = 2(x^2 - 3)$
       $x - 3y = 8$ $5x - 3y = 7$ $y - 2x = 18$ $2x - y = 2$

Diagnostic Test

Proof and Functions

These topics are covered on p.30-33.

22) Prove that the sum of any three consecutive odd numbers is a multiple of 3.

23) Naveen says, "for any integers x and y, $xy > y$". Prove that Naveen is wrong.

24) $f(x) = \dfrac{x+5}{3}$ and $g(x) = x - 3$.

   a) Evaluate $f(4)$.   b) Find $fg(x)$.   c) Find $f^{-1}(x)$.

Straight Lines and Quadratic Graphs

25) Give the gradient and y-intercept of the line $x + 2y = 4$.

26) Point A has coordinates $(5, 2)$ and point B has coordinates $(2, -4)$.
   a) Find the equation of the line passing through points A and B.
   b) Find the exact length of line AB.

Go to p.34-37 if you found these questions tricky.

27) Line A has equation $y = 2x + 5$.
   a) Find the equation of the line parallel to line A which passes through $(3, 2)$.
   b) Find the equation of the line perpendicular to line A which passes through $(2, 1)$.

28) Sketch the graph of $y = x^2 - 8x + 15$. Label the graph with the coordinates of the turning point and the points where the graph crosses the axes.

Harder Graphs and Graph Transformations

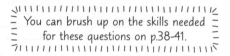

You can brush up on the skills needed for these questions on p.38-41.

29) Sketch the following graphs:

   a) $y = x^3$ b) $y = \dfrac{1}{x}$ c) $y = -\dfrac{1}{x}$

30) The graph on the right shows how the number of fish (F) living in a river changes over time. The equation of the graph is $F = mn^t$ where t is the number of years and m and n are positive constants. Find the values of m and n.

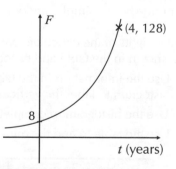

31) Find the equation of the tangent to $x^2 + y^2 = 25$ at the point $(3, 4)$. Give your answer in the form $ax + by + c = 0$.

32) $f(x) = x^2$. For parts a) to c) below, sketch the graphs of $y = f(x)$ and the given transformation.
   a) $y = f(x) + 3$ b) $y = f(x + 3)$ c) $y = -f(x)$

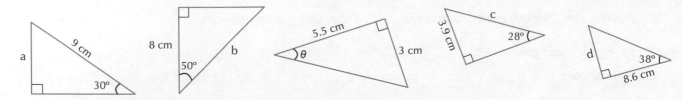

TEST YOURSELF **Diagnostic Test**

Trigonometry and Vectors

These topics are in Section 6 — p.42-50.

33) Find the unknowns in each of these triangles. Give your answers to 1 decimal place.

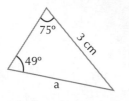 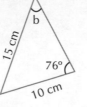 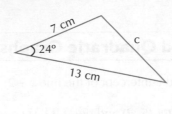 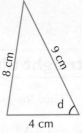

34) Find the unknowns in each of these triangles. Give your answers to 1 decimal place.

35) ABCD is the parallelogram shown on the right.
M, N, P and Q are the midpoints of the sides.
$\vec{AB} = \mathbf{a}$ and $\vec{BC} = \mathbf{b}$.
Find the following vectors in terms of $\mathbf{a}$ and $\mathbf{b}$.

a) $\vec{AC}$ b) $\vec{DQ}$ c) $\vec{CM}$

d) $\vec{QP}$ e) $\vec{MB}$ f) $\vec{PA}$

36) The diagram shows triangle ABC.
M is the midpoint of $\vec{AC}$ and N is the midpoint of $\vec{BC}$.
$\vec{AM} = 3\mathbf{a} - \mathbf{b}$ and $\vec{NC} = 2(\mathbf{a} - \mathbf{b})$.
Show that $\vec{AB}$ and $\vec{MN}$ are parallel.

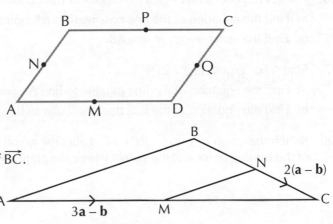

Sampling and Histograms

You'll find these topics on p.51-53.

37) Describe how a simple random sample of size 20 can be selected from a population of 200.

38) The weights of the chocolate bars in a shop storeroom
are shown in the table and histogram below.

a) Use the information in the table and the
histogram to label the vertical axis.

b) Use the histogram to complete the table.

c) Use the table to add the missing bar to the histogram.

Weight (w, in grams)	Frequency
$0 < w \le 100$	50
$100 < w \le 150$	100
$150 < w \le 200$	150
$200 < w \le 250$	

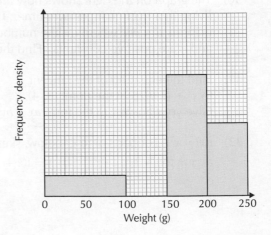

Diagnostic Test

Averages and Cumulative Frequency

Averages are covered on p.54-55.

39) Find the mean, median and mode(s) of these numbers:

5   3   –2   0   –3   2   1   1   4   2   6   11   –4

40) The table shows the journey times between
home and school for 60 students.

a) Write down the modal class.

b) Which group contains the median?

c) Estimate the mean value.

d) Draw a cumulative frequency graph
for the data in the table.

Time (m minutes)	Frequency
$5 < m \leq 10$	4
$10 < m \leq 15$	25
$15 < m \leq 20$	18
$20 < m \leq 25$	8
$25 < m \leq 30$	5

41) Using this cumulative frequency graph, find the:

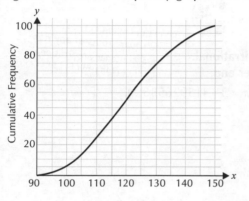

a) median

b) lower quartile

c) upper quartile

d) interquartile range

You can learn about cumulative frequency on p.56.

Probability and Tree Diagrams

You can learn about these topics on p.57-60.

42) Lewis asked 50 people if they like mashed potatoes (M) and roast potatoes (R).
The Venn diagram shows the results.

A person is chosen at random.
Find the probability that they:

a) like mashed potatoes

b) like neither mashed nor roast potatoes

c) like both types of potatoes

d) don't like roast potatoes

e) don't like mashed potatoes

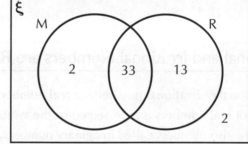

43) Mona's purse contains two £5 notes, four £10 notes and three £20 notes.
It also contains five 20p coins, four 50p coins and three £1 coins.

a) Mona picks one note and one coin at random from her purse.
Find the probability that she picks a £5 note and a 20p coin.

b) Mona picks two coins at random without replacement. Use a tree diagram
to find the probability she picks a 50p coin and a £1 coin.

Types of Number

First things first, here are some **important terms** that will crop up **throughout A-Level Maths**. If you're clear on what they mean now, it'll make A-Level life a lot easier.

Integers are just **Whole Numbers**

1) An **integer** is any positive or negative **whole number** (including **zero**).
2) The set of integers is represented by the symbol $\mathbb{Z}$.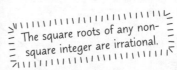

Alvin swore he wasn't asleep, he was just thinking about integers.

Rational Numbers can be Written as **Fractions**

1) A **rational** number is any number which can be written as a **fraction** with integers on the top and bottom. Don't forget, any **integer** can be written as a **fraction over 1**.
2) The set of **rational numbers** is represented by the symbol $\mathbb{Q}$.
3) Both **terminating** and **recurring decimals** are rational.
4) A number which **cannot** be written exactly as a fraction is **irrational**. Irrational numbers are **non-repeating** decimals, which **never end**.
5) For example, π is a **never-ending**, **non-repeating** decimal ($\pi = 3.1415926535...$). This means that π is **irrational**.

> The square roots of any non-square integer are irrational.

EXAMPLES

Is $0.\dot{6}$ a rational or irrational number?

$0.\dot{6}$ is the recurring decimal 0.666666... Recurring decimals are **rational** — $0.\dot{6} = \frac{2}{3}$.

Is 5.26 a rational or irrational number?

5.26 is **rational** because it is a terminating decimal $\left(5.26 = \frac{526}{100} = \frac{263}{50}\right)$.

All Rational and Irrational Numbers are **Real Numbers**

1) **Any rational** or **irrational** number is a **real number**.
2) The set of **real numbers** is represented by the symbol $\mathbb{R}$.
3) There are also numbers called **imaginary numbers**. They're **not real numbers**.
4) Imaginary numbers are the result of taking the **square root** of a **negative number**. It's **not possible** to do this with real numbers because the square of a real number is always positive, so **imaginary numbers** are used in these cases. You'll learn about them if you do A-Level Further Maths.

Rational? Irrational? Whatever man — I just keep it real...

1) Are these numbers rational or irrational? Explain your answer.
   a) 0.236849   b) 0.147$\dot{8}$ c) $\sqrt{64}$ d) $\sqrt{3}$ e) 2π
2) True or false?
   a) All integers are rational.   b) A recurring decimal is irrational.

Fractions

Multiplying Fractions — just Multiply the Numbers

1) Multiply the numerators and denominators **separately**.
2) Make sure to turn any **mixed numbers** into **improper fractions** first.
3) **Cancel** the fractions down **before** you start multiplying if you can — it'll make things easier.

EXAMPLE

Find $1\frac{3}{4} \times 2\frac{2}{3}$.

Write the mixed numbers as improper fractions:
Now cancel down across your fractions and multiply.

8 divides by 4, so you can cancel these down.

This is $1\frac{3}{4} = 1 + \frac{3}{4} = \frac{4}{4} + \frac{3}{4}$.

$$1\frac{3}{4} \times 2\frac{2}{3} = \frac{4+3}{4} \times \frac{6+2}{3} = \frac{7}{4} \times \frac{8}{3}$$

$$\frac{7}{{}_1\cancel{4}} \times \frac{\cancel{8}^2}{3} = \frac{7}{1} \times \frac{2}{3} = \frac{14}{3}$$

Dividing by a Fraction — Flip It and Multiply

1) To **divide** by a fraction, just turn the fraction you're dividing by **upside down**, and **change** the divide sign to a **multiply**.
2) Just like before, turn **mixed numbers** into **improper fractions** and do any **cancelling** before you start.

EXAMPLE

Calculate $\frac{3}{8} \div \frac{7}{16}$.

Flip the second fraction and multiply. Make sure you do any cancelling you can before multiplying. 16 divides by 8, so you can cancel these down:

$$\frac{3}{8} \div \frac{7}{16} = \frac{3}{\cancel{8}_1} \times \frac{\cancel{16}^2}{7} = \frac{3}{1} \times \frac{2}{7} = \frac{6}{7}$$

The Denominators need to Match for Addition and Subtraction

1) In order to **add** or **subtract** fractions, the **denominators** must be the **same**. So you have to find the **lowest common multiple** of all **denominators** (the lowest common denominator).
2) You still need to turn any mixed numbers into improper fractions.
3) Once both fractions have a common denominator, you can add/subtract their **numerators**.

EXAMPLE

Find $1\frac{5}{6} - \frac{3}{4}$.

You'll need to be confident with all of these rules as they're really important for using algebraic fractions.

Convert the mixed number to an improper fraction: $1\frac{5}{6} - \frac{3}{4} = \frac{11}{6} - \frac{3}{4}$.

Now they need a common denominator. The LCM of 6 and 4 is 12, so $\frac{11}{6} - \frac{3}{4} = \frac{22}{12} - \frac{9}{12}$.

This means you can now subtract the numerators: $\frac{22}{12} - \frac{9}{12} = \frac{13}{12}$.

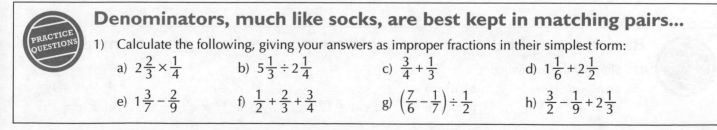

Denominators, much like socks, are best kept in matching pairs...

PRACTICE QUESTIONS

1) Calculate the following, giving your answers as improper fractions in their simplest form:

a) $2\frac{2}{3} \times \frac{1}{4}$

b) $5\frac{1}{3} \div 2\frac{1}{4}$

c) $\frac{3}{4} + \frac{1}{3}$

d) $1\frac{1}{6} + 2\frac{1}{2}$

e) $1\frac{3}{7} - \frac{2}{9}$

f) $\frac{1}{2} + \frac{2}{3} + \frac{3}{4}$

g) $\left(\frac{7}{6} - \frac{1}{7}\right) \div \frac{1}{2}$

h) $\frac{3}{2} - \frac{1}{9} + 2\frac{1}{3}$

Laws of Indices

Indices are Powers

For the value 4^3, 4 is the **base** and 3 is the **index**. (The plural of index is **indices**.)
You'll use indices all the time in A-Level Maths — especially for topics like **algebra**, **differentiation** and **integration**. So make sure you know the **laws of indices** inside out...

Multiplying Indices = Add the Powers, Dividing = Subtract them

So $\quad x^a \times x^b = x^{(a+b)} \quad$ and $\quad x^a \div x^b = x^{(a-b)}$

EXAMPLES

Simplify $a^4 \times a^2$.

Just add the indices: $a^4 \times a^2 = a^{(4+2)} = \mathbf{a^6}$.
You can see how this works by rewriting the calculation as one big multiplication:
$a^4 \times a^2 = (a \times a \times a \times a) \times (a \times a) = a \times a \times a \times a \times a \times a = a^6$

Rewrite $45 \div 43$ as a single power of 4.

This time, subtract the indices: $\qquad 4^5 \div 4^3 = 4^{(5-3)} = \mathbf{4^2}$
Again, you can see how this works by rewriting the calculation:
$4^5 \div 4^3 = \dfrac{\cancel{4} \times \cancel{4} \times \cancel{4} \times 4 \times 4}{\cancel{4} \times \cancel{4} \times \cancel{4}} = 4 \times 4 = 4^2$

Remember — this law only works when the two values have the same base. So $2^4 \times 2^7 = 2^{(4+7)} = 2^{11}$ is fine, but you can't use this rule to work out $2^4 \times 3^5$.

Simplify $8a^5 \times 2a^6$.

Both terms have the same base, a, and multiplication is commutative
(which means it doesn't matter what order you do it in).
So you can rewrite this as: $\ 8 \times a^5 \times 2 \times a^6 = (8 \times 2) \times (a^5 \times a^6) = 16 \times a^{(5+6)} = \mathbf{16a^{11}}$

Simplify $(x-1)^9 \div (x-1)^4 \times y^3 \times y^4$.

Deal with the powers of each base separately:
$(x-1)^9 \div (x-1)^4 \times y^3 \times y^4 = (x-1)^{(9-4)} \times y^{(3+4)} = \mathbf{(x-1)^5 y^7}$

Don't be put off by the brackets — $(x-1)$ is just the base.

There are Rules for x^0 and x^1

$x^0 = 1$ for any value of x The easiest way to see this is this with an example:
$4^3 \div 4^3 = 4^{(3-3)} = 4^0$ and $4^3 \div 4^3 = 1$, so $4^0 = 1$.

$x^1 = x$ for any value of x Again, this is easiest to understand as an example:
$4^3 \div 4^2 = 4^{(3-2)} = 4^1$, and $4^3 \div 4^2 = 64 \div 16 = 4$, so $4^1 = 4$.

PRACTICE QUESTIONS

Revise more I tell you — sorry these powers have gone to my head...

1) Simplify the following:

a) $b^5 \times b^6$

b) $a^9 \times a \times b^5$

c) $c^5 \div c^2$

d) $9y^{10} \div 3y^{-2}$

e) $x^2 \times x^3 \div x^4$

f) $z^3 \times (y+2)^5 \times z \div (y+2)^2$

g) $a^{(x+2)} \times a^{2x}$

h) $x^{-2}y^5 \times x^5 y^2$

Laws of Indices

To **Raise a Power** to **Another Power**, **Multiply** them

So $(x^a)^b = x^{ab}$

EXAMPLES

Simplify $(q^3)^2$.

$(q^3)^2 = q^{(3 \times 2)} = q^6$

> You can see how this works by rewriting it as one big multiplication:
> $(q^3)^2 = q^3 \times q^3 = (q \times q \times q) \times (q \times q \times q)$
> $= q \times q \times q \times q \times q \times q = q^6$

Express $2^6 \div 4^2$ as a single power.

You need to make the bases the same before simplifying, so rewrite 4 as a power of 2 using the rule above: $2^6 \div 4^2 = 2^6 \div (2^2)^2 = 2^6 \div 2^{(2 \times 2)} = 2^6 \div 2^4 = \mathbf{2^2}$

Negative and Fractional Indices are a bit Trickier

A **negative** index means '1 ÷ the positive power' — so $x^{-a} = \dfrac{1}{x^a}$

> For a fraction raised to a negative power, you turn the fraction upside down, then apply the positive index.

EXAMPLES

Write 3^{-4} as a fraction.

$3^{-4} = \dfrac{1}{3^4} = \dfrac{1}{81}$

Write $\dfrac{1}{125}$ as a power of 5.

$125 = 5^3$, so $\dfrac{1}{125} = \dfrac{1}{5^3} = 5^{-3}$

If a number has a **fractional** index, this means '**the root of**' — so $x^{\frac{1}{a}} = \sqrt[a]{x}$

EXAMPLES

Find $64^{\frac{1}{3}}$ without using a calculator.

$64^{\frac{1}{3}} = \sqrt[3]{64}$

This is the cube root of 64, so the answer is **4**.
(Because $4 \times 4 \times 4 = 64$.)

Find $\left(\dfrac{1}{256}\right)^{\frac{1}{4}}$ without using a calculator.

When you raise a fraction to a power, you raise the top and bottom to that power:

$\left(\dfrac{1}{256}\right)^{\frac{1}{4}} = \dfrac{1^{\frac{1}{4}}}{256^{\frac{1}{4}}} = \dfrac{\sqrt[4]{1}}{\sqrt[4]{256}} = \dfrac{1}{4}$

You might have to **rewrite** the index. Use the fact that $x^{\frac{a}{b}} = \left(x^{\frac{1}{b}}\right)^a$, which is the same as $(x^a)^{\frac{1}{b}}$.

If you get a **negative fractional index**, use this fact to turn $x^{-\frac{a}{b}}$ into $\left(x^{\frac{a}{b}}\right)^{-1}$ or $(x^{-1})^{\frac{a}{b}}$.

EXAMPLE

Find $27^{-\frac{1}{3}}$ without using a calculator.

$27^{-\frac{1}{3}} = \left(27^{\frac{1}{3}}\right)^{-1} = \left(\sqrt[3]{27}\right)^{-1} = (3)^{-1} = \dfrac{1}{3}$

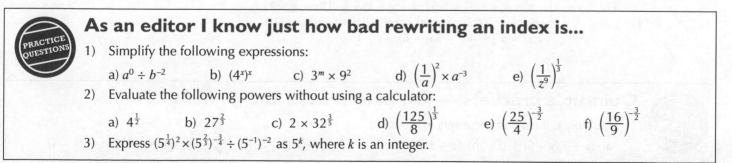

PRACTICE QUESTIONS

As an editor I know just how bad rewriting an index is...

1) Simplify the following expressions:

   a) $a^0 \div b^{-2}$ b) $(4^x)^x$ c) $3^m \times 9^2$ d) $\left(\dfrac{1}{a}\right)^2 \times a^{-3}$ e) $\left(\dfrac{1}{z^9}\right)^{\frac{1}{3}}$

2) Evaluate the following powers without using a calculator:

   a) $4^{\frac{1}{2}}$ b) $27^{\frac{2}{3}}$ c) $2 \times 32^{\frac{3}{5}}$ d) $\left(\dfrac{125}{8}\right)^{\frac{1}{3}}$ e) $\left(\dfrac{25}{4}\right)^{-\frac{3}{2}}$ f) $\left(\dfrac{16}{9}\right)^{-\frac{3}{2}}$

3) Express $(5^{\frac{1}{4}})^2 \times (5^{\frac{2}{3}})^{-\frac{3}{4}} \div (5^{-1})^{-2}$ as 5^k, where k is an integer.

Multiplying Out Brackets

Double Brackets and Squared Brackets Can Be Expanded

1) When expanding **double brackets**, multiply each term in one set of brackets by each term in the other. If each set of brackets contains **two** terms, you can use the **FOIL** method to make sure you don't miss any terms:

> Multiply the **First** term in each set of brackets together, then the two **Outside** terms, then the two **Inside** terms and finally the **Last** terms.

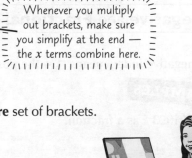
These techniques are really useful for lots of A-Level topics — like surds and solving various types of equations.

2) Write out **squared brackets** as double brackets and then use the FOIL method to avoid making silly mistakes.

Don't make the mistake of just squaring the stuff inside the brackets. In general, $(a + b)^2 \neq a^2 + b^2$.

EXAMPLE

Expand and simplify $(2x - 4)^2$.

$(2x - 4)^2 = (2x - 4)(2x - 4) = (2x \times 2x) + (2x \times -4) + (-4 \times 2x) + (-4 \times -4)$
$= 4x^2 - 8x - 8x + 16 = \mathbf{4x^2 - 16x + 16}$

Whenever you multiply out brackets, make sure you simplify at the end — the x terms combine here.

Use a Similar Method with Triple Brackets

1) You can think of **triple brackets** as **double brackets** multiplied by **one more** set of brackets.

2) So, use the **double bracket method** on two of the sets, then multiply the **result** by the **remaining set of brackets**.

3) You'll probably end up with **more than two terms** in your 'result', so you won't be able to use the FOIL method for the last bit. Instead, you should **break up** the **shorter** set of brackets so that each of its terms is multiplied by the **second** set of brackets **separately**.

4) Then multiply out each of the new sets of brackets **one at a time**.

EXAMPLE

Write $(x + 5)(x - 1)(x + 2)$ as a cubic expression.

$(x + 5)(x - 1)(x + 2) = (x + 5)(x^2 + x - 2) = x(x^2 + x - 2) + 5(x^2 + x - 2)$
$= (x^3 + x^2 - 2x) + (5x^2 + 5x - 10) = \mathbf{x^3 + 6x^2 + 3x - 10}$

Tina had mixed up 'cubic expressions' with 'cubist expressionism'.

Brackets can have More Than Two Terms

NEW CONTENT

At A-Level, you might be asked to multiply out sets of brackets with **more than two terms**. As usual, you need to multiply **every** term in the **first** set of brackets by **every** term in the **second**. The method for this is just the same as **stage two** of the **triple bracket method** above — multiply **each term** in the **shorter** set of brackets by the other set **separately**.

EXAMPLE

Expand and simplify the expression $(x^2 - 5x - 1)(x + y + 1)$.

$(x^2 - 5x - 1)(x + y + 1) = x^2(x + y + 1) + (-5x)(x + y + 1) + (-1)(x + y + 1)$
$= (x^3 + x^2y + x^2) + (-5x^2 - 5xy - 5x) + (-x - y - 1)$
$= \mathbf{x^3 + x^2y - 4x^2 - 5xy - 6x - y - 1}$

PRACTICE QUESTIONS

Quintuple brackets — just pack it all in and go home...

1) Multiply out each of these sets of brackets:
   a) $(y + 3)(y - 6)$ b) $(a - 3)(b + 4)$ c) $(p - 1)(p - 2q)^2$ d) $(s^2 + s + 2)(2s^2 - 2s + 4)$

Factorising

Factorising is Putting Brackets Into an Expression

Factorising means finding **common factors** that are in each term in an expression. You can take common factors outside a set of brackets to **rewrite** an expression. It's a handy skill that's used all the time in A-Level Maths, e.g. in **algebraic fractions** and **solving quadratics** and **cubics**.

1) You have to spot **common factors** in each term. These then come out to the **front** of the expression, and you put what's left in **brackets**.

2) Make sure that after you've finished factorising, any sets of brackets **can't be factorised** any more.

EXAMPLES

Factorise $14x + 21xy$.

$14x + 21xy = 7x(2 + 3y)$

Factorise $16x^3 + 4x^2 - 4x$.

$16x^3 + 4x^2 - 4x = 4x(4x^2 + x - 1)$

Check your answer by multiplying the brackets out again — if it's right, you'll end up back where you started...

At A-Level, you might see expressions where the common factor is a **set of brackets**. (NEW CONTENT)

EXAMPLE

Factorise $3x(x + 2) - 4(x + 2)$.

$(x + 2)$ appears in both terms of the expression, so it's a common factor:
$3x(x + 2) - 4(x + 2) = (x + 2)(3x - 4)$

The Difference Of Two Squares is One Square Minus Another

Any expression of the form $a^2 - b^2$ is called the **difference of two squares**.
It can be **factorised** really easily using this result:

$$a^2 - b^2 = (a + b)(a - b)$$

This result is used all the time to solve quadratics and cancel down algebraic fractions, so you need to be able to spot when you can use it.

You can see why this works by multiplying out $(a + b)(a - b)$:
$(a + b)(a - b) = a^2 - ab + ba - b^2 = a^2 - ab + ab - b^2 = a^2 - b^2$.

EXAMPLES

Factorise $x^2 - 25$.

$a = x$ and $b = 5$, so $x^2 - 25 = (x + 5)(x - 5)$

Factorise $4x^2 - 81$.

$4x^2 = (2x)^2$, so $4x^2 - 81 = (2x + 9)(2x - 9)$

If the number isn't a square, you can write it as a **square root squared** (see the next page).

EXAMPLE

Factorise $x^2 - 3y^2$.

$3y^2 = (\sqrt{3}\,y)^2$, so using the difference of two squares: $x^2 - 3y^2 = (x + \sqrt{3}\,y)(x - \sqrt{3}\,y)$.

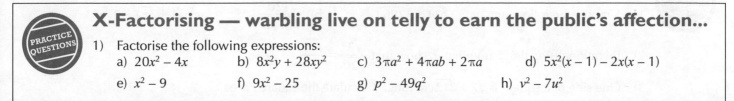

X-Factorising — warbling live on telly to earn the public's affection...

PRACTICE QUESTIONS

1) Factorise the following expressions:
   a) $20x^2 - 4x$
   b) $8x^2y + 28xy^2$
   c) $3\pi a^2 + 4\pi ab + 2\pi a$
   d) $5x^2(x - 1) - 2x(x - 1)$
   e) $x^2 - 9$
   f) $9x^2 - 25$
   g) $p^2 - 49q^2$
   h) $v^2 - 7u^2$

Surds

Surds are the Square Roots of Non-Square Numbers

Irrational numbers that can be written as **roots** ($\sqrt{}$) are called **surds.**
You need to be able to deal with them when working with **quadratics**, **cubics** and **vectors**.

There are Rules for Manipulating Surds

1) $\sqrt{a} \times \sqrt{b} = \sqrt{a \times b}$

Be careful — $\sqrt{a} + \sqrt{b}$ doesn't equal $\sqrt{a+b}$.

2) $a = (\sqrt{a})^2 = \sqrt{a}\sqrt{a}$

3) $\dfrac{\sqrt{a}}{\sqrt{b}} = \sqrt{\dfrac{a}{b}}$

This is just the difference of two squares with surds.

4) $(a + \sqrt{b})(a - \sqrt{b}) = a^2 - a\sqrt{b} + a\sqrt{b} - (\sqrt{b})^2 = a^2 - b$

EXAMPLES

Simplify $\sqrt{27} + 5\sqrt{3}$.

$$\sqrt{27} + 5\sqrt{3} = \sqrt{9 \times 3} + 5\sqrt{3}$$
$$= \sqrt{9} \times \sqrt{3} + 5\sqrt{3}$$
$$= 3\sqrt{3} + 5\sqrt{3} = 8\sqrt{3}$$

Expand and simplify $(2 + \sqrt{2})^2$.

Multiply out the brackets using FOIL:

$$(2 + \sqrt{2})^2 = (2 + \sqrt{2})(2 + \sqrt{2})$$
$$= 4 + 2\sqrt{2} + 2\sqrt{2} + \sqrt{2}\sqrt{2}$$
$$= 4 + 4\sqrt{2} + 2 = 6 + 4\sqrt{2}$$

Surds in the Denominator should be Rationalised

If there's a surd on the **bottom** of a fraction, you need to **get rid** of it. To do this, you multiply the **top** and **bottom** of the fraction by an expression that will give a **rational number** in the denominator. This is called **rationalising the denominator**.

EXAMPLES

Rationalise the denominator of $\dfrac{2\sqrt{2}}{\sqrt{5}}$.

You need to multiply top and bottom by something that will make the denominator rational —
$\sqrt{5}$ will work because $\sqrt{a}\sqrt{a} = a$:

$$\frac{2\sqrt{2}}{\sqrt{5}} = \frac{2\sqrt{2}\sqrt{5}}{\sqrt{5}\sqrt{5}} = \frac{2\sqrt{10}}{5}$$

Show that $\sqrt{80} + \dfrac{25}{\sqrt{5}} = 9\sqrt{5}$.

$\sqrt{80} = \sqrt{16}\sqrt{5} = 4\sqrt{5}$ and

$$\frac{25}{\sqrt{5}} = \frac{25\sqrt{5}}{\sqrt{5}\sqrt{5}} = \frac{25\sqrt{5}}{5} = 5\sqrt{5} \text{, so:}$$

$$\sqrt{80} + \frac{25}{\sqrt{5}} = 4\sqrt{5} + 5\sqrt{5} = 9\sqrt{5}$$

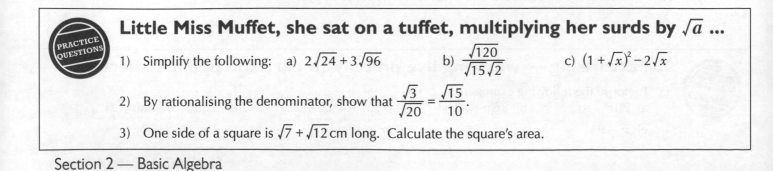

PRACTICE QUESTIONS

Little Miss Muffet, she sat on a tuffet, multiplying her surds by $\sqrt{a}$ ...

1) Simplify the following: a) $2\sqrt{24} + 3\sqrt{96}$ b) $\dfrac{\sqrt{120}}{\sqrt{15}\sqrt{2}}$ c) $(1 + \sqrt{x})^2 - 2\sqrt{x}$

2) By rationalising the denominator, show that $\dfrac{\sqrt{3}}{\sqrt{20}} = \dfrac{\sqrt{15}}{10}$.

3) One side of a square is $\sqrt{7} + \sqrt{12}$ cm long. Calculate the square's area.

Surds

You can **Rationalise** more **Complicated Denominators**

You'll need a slightly **different method** for rationalising more **complicated** denominators.

1) Just like before, you multiply the **top** and **bottom** of the fraction by the **same** expression.
2) The expression you need to multiply by is just the **denominator** with the **opposite sign** in front of the **surd**.

EXAMPLE

Rationalise the denominator of $\dfrac{3}{1+\sqrt{2}}$.

The bit you want to rationalise is $1+\sqrt{2}$, so multiply the top and bottom by $1-\sqrt{2}$:

$$\frac{3}{1+\sqrt{2}} = \frac{3(1-\sqrt{2})}{(1+\sqrt{2})(1-\sqrt{2})} = \frac{3-3\sqrt{2}}{1^2-(\sqrt{2})^2}$$
$$= \frac{3-3\sqrt{2}}{-1} = 3\sqrt{2}-3$$

This comes from the difference of two squares rule on the previous page: $(a+\sqrt{b})(a-\sqrt{b}) = a^2 - b$, with $a = 1$ and $b = 2$. You could also just multiply the brackets out.

This trick also works when there's a **number in front** of the **root** in the denominator. Just choose an expression with the **opposite sign** in front of the number multiplied by the surd.

EXAMPLE

Rationalise the denominator of $\dfrac{3+\sqrt{5}}{3+2\sqrt{5}}$.

Although this looks harder, just do the same thing.
The denominator is $3+2\sqrt{5}$, so multiply the top and bottom by $3-2\sqrt{5}$:

$$\frac{3+\sqrt{5}}{3+2\sqrt{5}} = \frac{(3+\sqrt{5})(3-2\sqrt{5})}{(3+2\sqrt{5})(3-2\sqrt{5})}$$
$$= \frac{3^2-6\sqrt{5}+3\sqrt{5}-2(\sqrt{5})^2}{3^2-6\sqrt{5}+6\sqrt{5}-(2\sqrt{5})^2}$$
$$= \frac{9-3\sqrt{5}-2\times 5}{9-2^2\times 5}$$
$$= \frac{-3\sqrt{5}-1}{-11} = \frac{3\sqrt{5}+1}{11}$$

Multiplying out these brackets is quite tricky, so take your time and don't skip any steps. See p.10 for more on multiplying out brackets.

Obasi was struggling to rationalise his fashion choices.

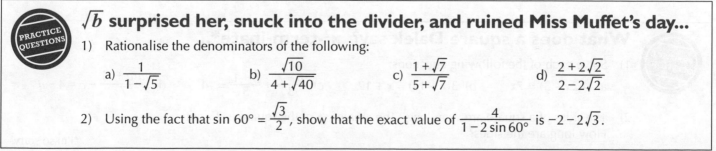

$\sqrt{b}$ surprised her, snuck into the divider, and ruined Miss Muffet's day...

1) Rationalise the denominators of the following:

   a) $\dfrac{1}{1-\sqrt{5}}$
   b) $\dfrac{\sqrt{10}}{4+\sqrt{40}}$
   c) $\dfrac{1+\sqrt{7}}{5+\sqrt{7}}$
   d) $\dfrac{2+2\sqrt{2}}{2-2\sqrt{2}}$

2) Using the fact that $\sin 60° = \dfrac{\sqrt{3}}{2}$, show that the exact value of $\dfrac{4}{1-2\sin 60°}$ is $-2-2\sqrt{3}$.

Solving Equations

Use the **6-Step Method** to **Solve Equations**

Lots of A-Level Maths requires you to be able to **solve equations**. The basic skills covered here crop up loads in **differentiation**, **trigonometry**, and solving **quadratics** and **cubics**.

To solve an equation, follow the **steps** below — you can **ignore** any steps you don't need. This method works for **any variable** — but if you're solving for x:

1) Remove any **fractions** by multiplying by the denominator(s).
2) **Multiply out** any **brackets** (see page 10).
3) **Collect** all the x-terms on one side and all the number terms on the other.
4) By **combining like terms**, **reduce** it to the form '$Ax = B$'.
5) Finally, **divide** both sides by **A**. This gives your answer '$x =$ '.
6) If you had '$x^2 =$ ' instead, **square root** both sides to end up with '$x = \pm$ '.

EXAMPLE

Solve $\dfrac{2x}{3} + \dfrac{4-x}{4} = 7x$.

1) First, get rid of the fractions. Multiply by 3 and then 4 to cancel the denominators:
$$\frac{3 \times 4 \times 2x}{3} + \frac{3 \times 4 \times (4-x)}{4} = 3 \times 4 \times 7x, \text{ so } 8x + 3(4-x) = 84x$$

2) Next, multiply out the set of brackets: $8x + 12 - 3x = 84x$

3) Now collect the x-terms on one side of the equation and the number terms on the other: $12 = 84x - 8x + 3x$

4) Combine like terms: $79x = 12$

5) And divide by 79 to get $x = \dfrac{12}{79}$

> You don't need step 6 because there's no x^2 term...

Here are a couple more examples — both have a **squared** term to deal with, so you'll need to use **step 6** of the method above.

Remember, when you take the **square root** of a number the answer can be **positive** or **negative**, so the equation will have **two solutions**.

> Be careful — the context of a question might mean only one of these answers will make sense.

EXAMPLES

Solve $3x(x + 4) = 12x + 6$.

Multiply out the brackets: $3x^2 + 12x = 12x + 6$

Simplify: $3x^2 + 12x - 12x = 6$
$$\Rightarrow 3x^2 = 6$$

Divide by 3: $x^2 = 2$

Take the square root: $x = \pm\sqrt{2}$

Solve $5a = \dfrac{125}{a}$.

Multiply by a to get rid of the fraction:
$5a^2 = 125$

Divide by 5: $a^2 = 25$

So, $a = \pm\sqrt{25} = \pm 5$

PRACTICE QUESTIONS

What does a square Dalek say? x^2-term-inate*...

1) Solve each of the following equations:

   a) $4(2x - 3) = 7x$ b) $3(x + 14) = x + 12$ c) $\dfrac{b-7}{3} + \dfrac{b+1}{5} = -1$ d) $\dfrac{q(q+7)}{7} - q = 4 - q^2$

2) The sides of a square are x cm long. The area of $\frac{1}{4}$ of the square is 25 cm². How long are the sides?

*I'm so sorry.

Rearranging Formulas

The **Subject** of a **Formula** is the Letter on its **Own**

1) If a formula has a **single letter** on one side of the equals sign, this letter is known as the **subject** of the formula. For example, y is the subject of the formula $y = mx + c$.

2) Sometimes you'll need to **rearrange** a formula to make a **different** letter the subject.

3) You'll use this a lot in the **mechanics** sections of A-Level Maths, especially with **constant acceleration formulas**, so make sure you've got it learned.

Use These Handy **7 Steps** to **Rearrange Formulas**

1) **Square** both sides to get rid of any **square roots**.

2) **Multiply** by the denominator(s) to get rid of any **fractions**.

3) **Multiply out** any **brackets**.

4) **Collect** all the **subject** terms together on one side and all **non-subject terms** on the other.

5) By **combining** like terms, **reduce** it to the form '**$Ax = B$**'. You might have to do some **factorising** to get it in the right form. ←

6) **Divide** both sides **by A** to give '$x =$'.

7) If you've got '$x^2 =$', **square root** both sides to get '$x = \pm$'.

This method is the same as the one for solving equations — so you've already had some practice at it.

Remember, x is the subject here. A and B could be numbers or letters (or a mix of both).

EXAMPLE

Make n the subject of the formula $\sqrt{\dfrac{n+3}{a}} = 2m$.

First, square both sides to get rid of the square root: $\dfrac{n+3}{a} = 4m^2$

Then multiply by a to get rid of the fraction: $n + 3 = 4am^2$

Finally, collect all the terms without an n on the RHS by subtracting 3: $n = 4am^2 - 3$

If the subject appears **more than once** in the formula, then you'll need to do some **factorising**.

EXAMPLE

Make a the subject of the formula $a^2 + b^2 = (3 + a)(a - b)$.

There aren't any square roots or fractions, so go straight to multiplying out the brackets: $a^2 + b^2 = 3a - 3b + a^2 - ab$.

Gather all the terms containing a on one side, and everything else on the other side: $a^2 - 3a - a^2 + ab = -3b - b^2$.

This can be simplified to $-3a + ab = -3b - b^2$.

Take out the common factors a on the LHS and $-b$ on the RHS to give: $a(b - 3) = -b(b + 3)$

Finish off by dividing both sides by $(b - 3)$ to get $a = \dfrac{-b(b+3)}{b-3}$

$(b - 3)$ is the same as $(-3 + b)$, just rewritten because it's neater.

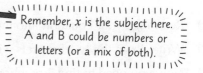

Loyal subjects I command thee — moveth to yonder side of the '='...

PRACTICE QUESTIONS

1) a) Express a in terms of b, given that $b(a + 2) = 4$. 　　b) Make c the subject of $f = \dfrac{9}{5}c + 32$.

2) a) Make x the subject of the formula $y = \dfrac{2x^2 - 3}{x^2 - 1}$. 　　b) Make t the subject of $s = \dfrac{\sqrt{t+u}}{u}$

Factorising Quadratics

Quadratics have a Specific Form

Quadratics are of the form $y = ax^2 + bx + c$, where a, b and c are **numbers**.

A quadratic equation can't have any other powers of x, like x^3 or $x^{\frac{1}{2}}$.

You'll need to be really confident with quadratics for A-level — they crop up **everywhere**, including in later work on **polynomials** and **graphs** of functions.

> 'a' and 'b' are the coefficients of x^2 and x. 'c' is called the constant term.

Factorising Quadratics is Easier when a = 1

1) Normally, you'll factorise your **quadratic** by splitting it up into **two** sets of **brackets** — $x^2 + bx + c = (x + p)(x + q)$.

2) As well as factorising a quadratic, you might be asked to **solve** the equation. This just means finding the **values** of x that make each **bracket** equal 0 (see example below).

3) Factorising quadratics is easier when the coefficient of x^2 is 1.

---

1) Rearrange your equation into the **standard format**: $x^2 + bx + c = 0$

2) Write down the **two brackets** with the x's in: $(x \quad)(x \quad) = 0$

3) Now you need to find 2 numbers that **multiply** together to give the value 'c' (the constant) but also **add/subtract** to give the value 'b' (the coefficient of x).

4) Put in the +/– signs and make sure they give the right numbers.

5) **Check** this works by **expanding the brackets** to make sure this gives you the original equation.

6) To solve the equation, set each **bracket equal to 0** and solve for x.

> This first step might involve multiplying or dividing. E.g. dividing $3x^2 + 6x + 3 = 0$ by 3 gives $x^2 + 2x + 1 = 0$.

> For more on expanding brackets, see page 10.

EXAMPLE

Solve $x^2 + 7x + 12 = 0$.

This is already in the standard format, with $b = 7$ and $c = 12$.

Write out the initial brackets: $(x \quad)(x \quad) = 0$.

You need to find a pair of numbers that multiply to give c (= 12) and add together to give b (= 7): $3 \times 4 = 12$ and $3 + 4 = 7$ — perfect.

Write these in the brackets, and then fill in the +/– to make 3 and 4 add to give 7: $(x + 3)(x + 4) = 0$.

Double check that this works by multiplying out the brackets
$(x + 3)(x + 4) = x2 + 4x + 3x + 12 = x2 + 7x + 12$ ✔

> If c is positive, then both signs are the same. If c is negative, then the signs will be opposite.

Set the brackets equal to 0 to solve: $(x + 3) = 0$ so $x = -3$ and $(x + 4) = 0$ so $x = -4$

What do you call a group of equations? A squadratic...

> PRACTICE QUESTIONS

1) Solve the following quadratic equations:
   a) $x^2 + 3x + 2 = 0$
   b) $x^2 + 8x + 7 = 0$
   c) $x^2 = 2(7x - 20)$
   d) $x^2 - x = 6$
   e) $7x = x^2 + 10$
   f) $\frac{x^2}{2} + 2x - 6 = 0$
   g) $x - 4 - \frac{12}{x} = 0$
   h) $2x^2 - 2x - 4 = 0$

2) Factorise $x^2 - 2xz + z^2$.

3) In an experiment, the temperature $T \,°C$ is modelled by $T = -m^2 + 13m - 30$, where m is the time in minutes after the start of the experiment. Find both times at which the temperature is $0 \,°C$.

Factorising Quadratics

Factorising a **Quadratic** is **Trickier** when **a ≠ 1**

When $a \neq 1$, the basic method is still the same as before, but it's just a bit more **fiddly**.

1) When you write down the two brackets, they'll now be of the form $(nx \quad)(mx \quad)$, where n and m are two numbers that **multiply** to give a.
2) Now you're looking for two numbers that **multiply** to give c, but which also give you bx when you **multiply** them by nx and mx, and then **add / subtract** them.
3) Put those two numbers in the brackets, and then choose their **signs** to make it work.

nx, mx, bx... bmx?

EXAMPLES

Solve $2x^2 + 7x = -6$.

First rearrange the equation into the standard format: $2x^2 + 7x + 6 = 0$

Now, write out the two brackets. The x bits need to multiply together to be $2x^2$ so the brackets will be of the form $(2x \quad)(x \quad)$.

Now, find pairs of numbers that multiply to give c (= 6).
Number pairs 1×6 and 2×3 both work.

Try these number pairs out in the brackets until you find a pair that gives $7x$.
You'll need to try each number in 2 positions because the brackets are different.
$(2x \quad 1)(x \quad 6)$ multiplies to give $12x$ and x which add/subtract to give $13x$ or $11x$. ✗
$(2x \quad 6)(x \quad 1)$ multiplies to give $2x$ and $6x$ which add/subtract to give $8x$ or $4x$. ✗
$(2x \quad 2)(x \quad 3)$ multiplies to give $6x$ and $2x$ which add/subtract to give $8x$ or $4x$. ✗
$(2x \quad 3)(x \quad 2)$ multiplies to give $4x$ and $3x$, which add/subtract to give x or $7x$. ✔

This means you can fill in the +/– signs so that 4 and 3 add to give 7: $(2x + 3)(x + 2)$.

Check that this works by multiplying it out:
$(2x + 3)(x + 2) = 2x^2 + 4x + 3x + 6 = 2x^2 + 7x + 6$ ✔

Set the brackets equal to zero and solve: $(2x + 3) = 0$, so $x = -\frac{3}{2}$ and $(x + 2) = 0$, so $x = -2$

Factorise $3x^2 + 4x - 7$.

The brackets will be $(3x \quad)(x \quad)$.

The only possible number pairing is 1 and 7, so try each number in both positions:
$(3x \quad 1)(x \quad 7)$ multiplies to give $21x$ and x which add/subtract to give $22x$ or $20x$.
$(3x \quad 7)(x \quad 1)$ multiplies to give $3x$ and $7x$ which add/subtract to give $10x$ or $4x$. ✔

Finally work out the signs for each set of brackets: $3x^2 + 4x - 7 = (3x + 7)(x - 1)$.

Don't forget to multiply out the brackets to check your answer is correct.

Oh drat... Double drat.... Triple drat... Quad drat it...

1) Solve the following quadratic equations:
   a) $2x^2 + 9x + 9 = 0$ b) $5x^2 + 13x + 6 = 0$ c) $2x^2 = x + 10$ d) $3x + \frac{21}{x} = 16$

2) The equation $h = -4t^2 + 2t + 2$ models the height h m a paper plane is from the ground, where t is time after the plane is thrown in seconds. At what time t does the plane hit the ground?

The Quadratic Formula

The **Quadratic Formula** Gives Solutions to **Quadratic Equations**

The solutions to **any** quadratic equation in the form
$ax^2 + bx + c = 0$ are given by this formula:

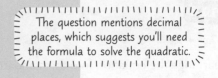

$$x = \frac{-b \pm \sqrt{b^2 - 4ac}}{2a}$$

It means you can solve quadratics that are too hard to factorise.
1) To find x, you just need to put the values of a, b and c into the formula.
2) The ± sign means you can end up with two solutions —
   in the final stage you replace it with '+' and then '–'.

EXAMPLE

Solve $8x^2 + 3x - 3 = 0$, giving your answers to 3 decimal places.

$a = 8$, $b = 3$ and $c = -3$. So put these values into the formula:

$$x = \frac{-3 \pm \sqrt{3^2 - 4 \times 8 \times -3}}{2 \times 8} = \frac{-3 \pm \sqrt{9 + 96}}{16} = \frac{-3 \pm \sqrt{105}}{16}$$

> The question mentions decimal places, which suggests you'll need the formula to solve the quadratic.

So $x = \dfrac{-3 + \sqrt{105}}{16}$ and $x = \dfrac{-3 - \sqrt{105}}{16}$ are the solutions —

you should leave your answers in this form if you're asked for exact solutions.

Put these into your calculator to get: $x = 0.453$ and $x = -0.828$

> At A-Level, you might see quadratics which result in negative square roots in the quadratic formula (see below).

The **Discriminant** tells you about the **Roots** of a **Quadratic**

The **discriminant** is the bit of the quadratic formula that's **inside** the **square root**: $b^2 - 4ac$
It's used in A-Level Maths to find **how many roots** a quadratic has.
The roots are the **values of x** where the **graph** of the quadratic **crosses the x-axis**.

1) If the discriminant is **positive** ($b^2 - 4ac > 0$), this means the quadratic has
   two real roots. The formula will give **two different real number values**
   of x — one from adding and one from subtracting the $\sqrt{b^2 - 4ac}$ bit.
2) If the discriminant is **zero**, the quadratic only has **one root** because
   adding or subtracting 0 will give the same value of x.
3) If the discriminant is **negative** ($b^2 - 4ac < 0$) this means the quadratic has **no (real) roots**
   because the square root of a negative number is **not** a real number (see page 6).

EXAMPLE

Find the discriminant of $4x^2 + 3x + 1$.
How many real roots does $4x^2 + 3x + 1 = 0$ have?

$a = 4$, $b = 3$ and $c = 1$, so the discriminant is $b^2 - 4ac = 3^2 - 4 \times 4 \times 1 = 9 - 16 = -7$

The discriminant is < 0, so $4x^2 + 3x + 1 = 0$ has **no real roots**.

Due to the poor quality of previous jokes, this one's been removed...

1) Solve the following quadratic equations. Give your answers to 3 d.p.
   a) $4x^2 - 7x + 1 = 0$ b) $6x^2 + x = 4$ c) $-2x^2 = 3x - 4$
2) Find the discriminant of $5x^2 + 7x + 3$. How many real roots does $5x^2 + 7x + 3 = 0$ have?

Completing the Square

Here's How to **Complete The Square** when **a = 1**

Completing the square means rewriting a **quadratic** as a **squared bracket** plus or minus a **number**. It's used at A-Level for **sketching graphs** (see next page) and **solving the equations** of quadratics (see below) and it's also used with the **equations of circles**.
This is the method for when a (the coefficient of x^2) is 1:

1) Start by getting your quadratic into the **standard format**: $ax^2 + bx + c$.
2) Write out the initial **squared bracket** $(x + \frac{b}{2})^2$ — so just **divide** the coefficient of x by **2**.
3) Now, **multiply out** the squared bracket and **compare** what you get with the **original** quadratic to see what you need to add or subtract.
4) Finally, add or subtract the **adjusting number** to make it **match** the original.

EXAMPLE

Rewrite $x^2 + 12x + 16$ in the form $(x + m)^2 + n$.

This quadratic's already in the standard format, so go straight to step 2 and write out the initial squared bracket: $(x + 6)^2$

Multiply this out and compare with the original: $(x + 6)^2 = x^2 + 12x + 36$

This has a 36 at the end, but you need a 16 at the end, so adjust by $16 - 36 = -20$

$(x + 6)^2 - 20 = x^2 + 12x + 36 - 20 = x^2 + 12x + 16$ ✔

So the completed square is $(x + 6)^2 - 20$

> To find the adjusting number, you can just take the number term you get from expanding the bracket away from 'c'.

The **Completed Square** can help you **Solve** a **Quadratic**

Once you've **completed the square**, you can **solve** a quadratic. All you have to do is **rearrange** the completed square to make x the **subject** by following these three steps:

1) Move the **adjusting number** to the **other side** of the equation.
2) **Square root** both sides — remember the '±'...
3) Rearrange to get x on its **own**.

EXAMPLE

Given that $x^2 - 14x + 36 = (x - 7)^2 - 13$, solve $x^2 - 14x + 36 = 0$.

Set the completed square equal to zero to solve $(x - 7)^2 - 13 = 0$

Now, move the adjusting number to the RHS: $(x - 7)^2 = 13$

Next, square root both sides to get: $x - 7 = \pm\sqrt{13}$

Finally, add seven to each side to get x on its own: $x = 7 + \sqrt{13}$ and $x = 7 - \sqrt{13}$

> Remember, when you take a square root you get two values — one positive and one negative.

PRACTICE QUESTIONS

Complete Trafalgar Square — 1 Column + 4 plinths + 5^{10} pigeons...

1) Write each of the following in the form $(x + m)^2 + n$:
   a) $x^2 + 4x + 1$ b) $x^2 - 12x + 5$ c) $x^2 - 20x - 10$ d) $x^2 + 7x - 3$
2) Write $x^2 + 6x - 8$ in the form $(x + m)^2 + n$. Hence solve $x^2 + 6x - 8 = 0$.

Completing the Square

You can **Sketch** the **Graph** Using the **Completed Square**

The completed square form of a quadratic gives you information about its graph.
When the coefficient of x^2 is **positive**:

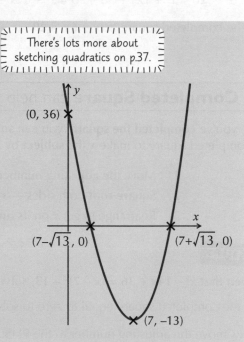

> If the adjusting number is O, the equation has one root.

- The graph is **u-shaped** with a **minimum point** (see p.37).
- The **adjusting number** tells you the y-value of this minimum point.
  This is because the **smallest value** the quadratic can take occurs when the bit in the brackets is **zero** (the bracket is squared so can never give a negative value).
- If the **adjusting number** is **positive** then the graph's **minimum point** is **above** the x-axis. So the graph **never** crosses the x-axis — it has **no real roots**.

When the coefficient of x^2 is **negative** (there are examples where a isn't 1 on the next page):

- The graph is **n-shaped** with a **maximum point** (see p.37).
- The **maximum point** occurs when the bit in brackets is **zero** and the **adjusting number** tells you the y-value of this maximum point.
- If the **adjusting number** is **negative** then the graph's **maximum point** is **below** the x-axis. So the graph **never** crosses the x-axis, meaning it has **no real roots**.

As you saw on the last page, the completed square can also be used to find the solutions of the quadratic. So it can tell you where the graph **crosses the x-axis** (if it crosses the x-axis at all).

EXAMPLE

Sketch the graph of $y = x^2 - 14x + 36$.

You know from the example on the previous page that
$x^2 - 14x + 36 = (x - 7)^2 - 13$.

The coefficient of x^2 is positive, so it is
u-shaped and has a minimum point.

The graph takes its minimum value
when $(x - 7)^2 = 0$, so when $x = 7$.
The minimum y-value is the adjusting number (-13).

So the minimum point of the graph is at $(7, -13)$.

The graph crosses the x-axis at the solutions to the
quadratic equation. From the example on the previous page,
you know these solutions are $x = 7 + \sqrt{13}$ and $x = 7 - \sqrt{13}$.

To find where the graph crosses the y-axis, let $x = 0$:
$0^2 - (14 \times 0) + 36 = 36$, so it crosses at $(0, 36)$.

> There's lots more about sketching quadratics on p.37.

$(0, 36)$
$(7-\sqrt{13}, 0)$ $(7+\sqrt{13}, 0)$
$(7, -13)$

Complete Times Square — Billboards + billboards + billboards + billboards + ...

1) a) Write $x^2 + 16x + 3$ in the form $(x + a)^2 + b$.
   b) Hence solve $x^2 + 16x + 3 = 0$.
   c) Use this information to sketch the graph of $y = x^2 + 16x + 3$.

2) The graph on the right shows the graph of $y = (x + p)^2 + q$,
   where p and q are integers.
   a) Use the sketch to find the values of p and q.
   b) Find the y-intercept of the graph.

Completing the Square

Here's How to **Complete The Square** where a ≠ 1

If *a* isn't 1, you need to take a **factor** of *a* from the x^2 and *x* terms before you can complete the square.

This works for negative or fractional values of 'a' as well.

EXAMPLE

Write $3x^2 - 8x = -4$ in the form $p(x + q)^2 + r = 0$.

Start by putting the quadratic into the standard format: $3x^2 - 8x + 4 = 0$.

Take out a factor of 3 from the x^2 and *x* terms: $3(x^2 - \frac{8}{3}x) + 4 = 0$.

Now, write out the initial bracket: $3(x - \frac{8}{6})^2 = 3(x - \frac{4}{3})^2$

Multiply out the bracket and compare to the original: $3(x - \frac{4}{3})^2 = 3x^2 - 8x + \frac{16}{3}$

The adjusting number is $4 - \frac{16}{3} = -\frac{4}{3}$:
$3(x - \frac{4}{3})^2 - \frac{4}{3} = 3x^2 - 8x + \frac{16}{3} - \frac{4}{3} = 3x^2 - 8x + 4$ ✔

So the completed square is $3(x - \frac{4}{3})^2 - \frac{4}{3}$

Completing the Square is useful for **Circle Equations**

NEW CONTENT

At A-Level, you'll sometimes need to rearrange the **equation of a circle** so you can find its **centre** and **radius**. There's lots more about the equations of circles on page 39. With circles you might need to complete the square **twice** — once for *x* and once for *y*.

EXAMPLE

Write $x^2 + y^2 - 6x + 14y + 22 = 0$ in the form $(x - m)^2 + (y - n)^2 = k$.

Start by rearranging the equation so the *x*'s and *y*'s are grouped together:
$x^2 - 6x + y^2 + 14y + 22 = 0$

Now you need to complete the square for both *x* and *y*.
Completing the square for $x^2 - 6x$ gives $(x - 3)^2 - 9$,
so the equation so far is: $(x - 3)^2 - 9 + y^2 + 14y + 22 = 0$

Completing the square for $y^2 + 14y$ gives: $(y + 7)^2 - 49$,
so the equation is: $(x - 3)^2 - 9 + (y + 7)^2 - 49 + 22 = 0$

Finally, rearrange into the form $(x - m)^2 + (y - n)^2 = k$:
Combine all the number terms together: $(x - 3)^2 + (y + 7)^2 - 36 = 0$
Put the number term on the other side of the equation: $(x - 3)^2 + (y + 7)^2 = 36$

Barney and Pip were meant to be completing the square, but they kept going round in circles.

Complete Red Square — Kremlin + fine this joke's getting old now...

PRACTICE QUESTIONS

1) Complete the square for each of these quadratics:
   a) $2x^2 - 3x - 10$ b) $-x^2 - 3x - 1$ c) $5x^2 - x - 3$ d) $-\frac{1}{2}x^2 - 3x - 7$
2) Write $x^2 + y^2 - 2x + 6y - 15 = 0$ in the form $(x - m)^2 + (y - n)^2 = k$.

Algebraic Fractions

Simplify Algebraic Fractions by Cancelling

To simplify an **algebraic fraction**, you need to find **common factors** in the numerator and denominator. Then you can **divide** the numerator and denominator by the common factors to cancel them. Some fractions won't cancel straight away — you'll need to **factorise** first.

EXAMPLES

Simplify $\dfrac{36ab^5}{27a^2b^3}$.

The common factors in the numerator and denominator are 9, a and b^3.
Cancel down by dividing the numerator and denominator by each one: $\dfrac{\overset{4}{\cancel{36}}a\cancel{b^5}^{b^2}}{\underset{3}{\cancel{27}}\cancel{a^2}_{a}\cancel{b^3}} = \dfrac{4b^2}{3a}$

Simplify $\dfrac{8x^3 - 4x^2}{6x^2 + 5x - 4}$.

Factorising the expressions in the numerator and denominator gives: $\dfrac{4x^2(2x-1)}{(2x-1)(3x+4)}$

Then cancelling $(2x - 1)$ gives: $\dfrac{4x^2\cancel{(2x-1)}}{\cancel{(2x-1)}(3x+4)} = \dfrac{4x^2}{3x+4}$

Take a look back at p.16-17 to recap factorising quadratics.

Multiplying and Dividing Algebraic Fractions

To **multiply** algebraic fractions, multiply the numerators and denominators **separately**.
To **divide** algebraic fractions, turn the second fraction **upside down** and then **multiply**.

EXAMPLES

Always do as much cancelling down as you can before multiplying.

Simplify $\dfrac{x+3}{2x} \times \dfrac{x^2}{(x+3)^2}$.

Cancel any common factors first: $\dfrac{\cancel{x+3}}{2x} \times \dfrac{\cancel{x^2}^{x}}{(x+3)^{\cancel{2}}}$

Then multiply the numerator and denominator separately: $\dfrac{x}{2(x+3)}$

You've seen this method with numbers on page 7.

Simplify $\dfrac{x^2 - 16}{8} \div \dfrac{(x-4)}{2x}$.

Start by flipping the second fraction upside down to give: $\dfrac{x^2 - 16}{8} \times \dfrac{2x}{(x-4)}$

You can factorise $x^2 - 16$ using the rule for a difference of two squares — see p.11.

Factorise and cancel the common factors $(x - 4)$ and 2: $\dfrac{\cancel{(x-4)}(x+4)}{\underset{4}{\cancel{8}}} \times \dfrac{\overset{}{\cancel{2}x}}{\cancel{(x-4)}}$

Then multiply to get the answer: $\dfrac{x(x+4)}{4}$

PRACTICE QUESTIONS

Flipping, factorising, multiplying — so much to do, so little time...

1) Simplify these fractions:

   a) $\dfrac{3x^2}{7x}$ b) $\dfrac{8x+16}{2x-4}$ c) $\dfrac{x^2-25}{5(x+5)}$ d) $\dfrac{3x^2+16x-12}{2x^2+13x+6}$ e) $\dfrac{x^3+2x^2+x}{x^2-3x-4}$

2) Simplify these expressions:

   a) $\dfrac{x+3}{x^2} \times \dfrac{x}{4}$ b) $\dfrac{3x+9}{4} \times \dfrac{x}{3(x+3)}$ c) $\dfrac{10x}{x^2-9} \div \dfrac{2x+14}{x-3}$ d) $\dfrac{x^2+5x+6}{9} \div \dfrac{x^2-4x-21}{6x-42}$

Algebraic Fractions

Adding and Subtracting Algebraic Fractions is a Little Tougher

Before you add or subtract algebraic fractions, take a look at the **denominators**.
If they're all the **same**, they have a **common denominator**,
so you can just add up the numerators.

$$\frac{2}{x} + \frac{5a}{x} + \frac{2a^2}{x} = \frac{2 + 5a + 2a^2}{x}$$

But when the denominators are **different**, you'll need to:

1) Find a **common denominator** — this should be the **lowest common multiple** of all the denominators.
2) **Rewrite** each fraction with the common denominator by **multiplying** the top and bottom by the **same thing**.
3) Then make into one fraction by **adding** or **subtracting** the **numerators**.

These skills are really useful — algebraic fractions pop up in both years of A-Level.

EXAMPLES

Simplify $\dfrac{3x+y}{2} - \dfrac{x-2y}{7}$.

The LCM of 2 and 7 is 14, so rewrite each fraction with a denominator of 14.
Multiply the top and bottom of the first fraction by 7,
and the top and bottom of the second fraction by 2:

$$\frac{3x+y}{2} - \frac{x-2y}{7} = \frac{7(3x+y)}{14} - \frac{2(x-2y)}{14} = \frac{21x + 7y - 2x + 4y}{14} = \mathbf{\frac{19x + 11y}{14}}$$

Simplify $\dfrac{1}{x} + \dfrac{3}{x+1}$.

Multiply x and $(x + 1)$ together to get the common denominator $x(x + 1)$.
Rewrite each fraction with a denominator of $x(x + 1)$ to get:

$$\frac{1}{x} + \frac{3}{x+1} = \frac{1(x+1)}{x(x+1)} + \frac{3x}{x(x+1)} = \frac{x+1+3x}{x(x+1)} = \mathbf{\frac{4x+1}{x(x+1)}}$$

You can use exactly the same method to add and subtract **more than two** fractions.

EXAMPLE

Simplify $\dfrac{1}{4} + \dfrac{1}{2x} - \dfrac{1}{4x}$.

The LCM of the three denominators is $4x$ — so this is the simplest common denominator.

Rewriting each fraction over $4x$ gives: $\dfrac{x}{4x} + \dfrac{2}{4x} - \dfrac{1}{4x} = \mathbf{\dfrac{x+1}{4x}}$

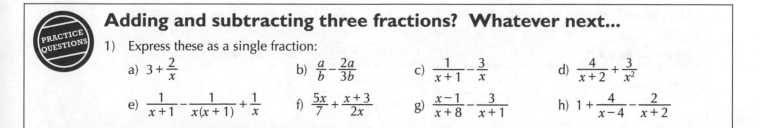

PRACTICE QUESTIONS

Adding and subtracting three fractions? Whatever next...

1) Express these as a single fraction:

a) $3 + \dfrac{2}{x}$

b) $\dfrac{a}{b} - \dfrac{2a}{3b}$

c) $\dfrac{1}{x+1} - \dfrac{3}{x}$

d) $\dfrac{4}{x+2} + \dfrac{3}{x^2}$

e) $\dfrac{1}{x+1} - \dfrac{1}{x(x+1)} + \dfrac{1}{x}$

f) $\dfrac{5x}{7} + \dfrac{x+3}{2x}$

g) $\dfrac{x-1}{x+8} - \dfrac{3}{x+1}$

h) $1 + \dfrac{4}{x-4} - \dfrac{2}{x+2}$

Inequalities

Solving **Inequalities** is Just Like Solving **Equations**...

...whatever you do to one side, you also do to the other side. And follow this important rule:

> If you **multiply** or **divide** by a **negative number**, **flip** the inequality sign.

Inequalities in the form $ax + b > cx + d$ are known as linear inequalities.

EXAMPLES

Solve $8x < 5x - 3$.

Subtract $5x$ from both sides:
$8x - 5x < 5x - 3 - 5x$
$\quad 3x < -3$

Then divide by 3:
$3x \div 3 < -3 \div 3$
$\quad \mathbf{x < -1}$

Solve $3(6 - x) \geq 12$.

Multiply out the brackets first: $18 - 3x \geq 12$

Subtract 18 from both sides:
$18 - 3x - 18 \geq 12 - 18$
$\quad\quad -3x \geq -6$

Divide by -3:
$-3x \div -3 \leq -6 \div -3$
$\quad\quad \mathbf{x \leq 2}$

You've divided by a negative number, so flip the sign.

At A-Level, you might need to find values that satisfy **two** inequalities...

EXAMPLE

NEW CONTENT

Find the values of x that satisfy both $3x + 4 > 2x - 1$ and $4x \geq 8x + 12$.

Solve each inequality first:

$$3x + 4 > 2x - 1 \qquad\qquad 4x \geq 8x + 12$$
$$x + 4 > -1 \qquad\qquad\quad -4x \geq 12$$
$$x > -5 \qquad\qquad\qquad x \leq -3$$

Draw a number line and show both solutions:

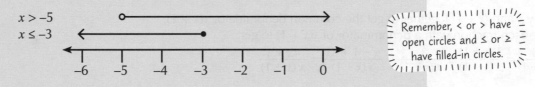

$x > -5$
$x \leq -3$

Remember, < or > have open circles and ≤ or ≥ have filled-in circles.

The set of values where the lines overlap satisfy both inequalities, so: $\mathbf{-5 < x \leq -3}$

Quadratic Inequalities are a bit Different

1) **Quadratic inequalities** have a **squared term** in them — like quadratic equations.
2) Quadratic equations can have two solutions (see p.16). Quadratic inequalities are similar — their solutions can be **two sets of values**, or a **range enclosed by two values**.

EXAMPLES

Solve $3x^2 - 22 > 5$.

Add 22 to both sides: $3x^2 - 22 + 22 > 5 + 22$
$\quad\quad\quad\quad\quad\quad\quad 3x^2 > 27$

Then divide both sides by 3: $3x^2 \div 3 > 27 \div 3$
$\quad\quad\quad\quad\quad\quad\quad\quad\quad x^2 > 9$

So $\mathbf{x < -3}$ or $\mathbf{x > 3}$

You can check your solutions have the right inequality signs by trying some values. E.g. to check $x < -3$, try $x = -4$: $x^2 = 16$ which is > 9, so < is correct.

Solve $4x^2 - 14 \leq 2$.

Add 14 to both sides: $4x^2 - 14 + 14 \leq 2 + 14$
$\quad\quad\quad\quad\quad\quad\quad 4x^2 \leq 16$

Divide both sides by 4: $4x^2 \div 4 \leq 16 \div 4$
$\quad\quad\quad\quad\quad\quad\quad\quad x^2 \leq 4$

So $\mathbf{-2 \leq x \leq 2}$

Inequalities

Sketch a Graph to Help You Solve Quadratic Inequalities

When a **quadratic inequality** also has **x-terms** (e.g. $ax^2 + bx + c < d$), it's a bit trickier
to solve. The best method is to turn the inequality into an **equation** you can **sketch**
— see p.37 for more on sketching quadratics. Then you can solve
the inequality by looking for the part of the graph that satisfies it.

EXAMPLE

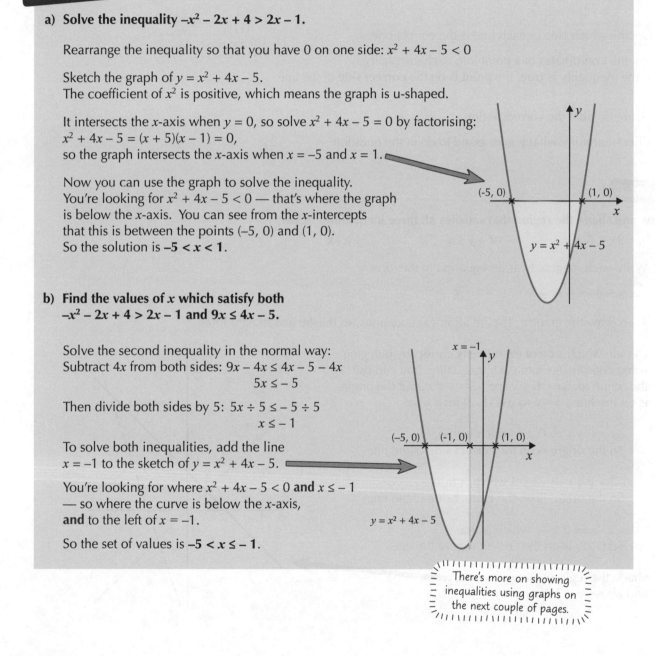

a) **Solve the inequality $-x^2 - 2x + 4 > 2x - 1$.**

Rearrange the inequality so that you have 0 on one side: $x^2 + 4x - 5 < 0$

Sketch the graph of $y = x^2 + 4x - 5$.
The coefficient of x^2 is positive, which means the graph is u-shaped.

It intersects the x-axis when $y = 0$, so solve $x^2 + 4x - 5 = 0$ by factorising:
$x^2 + 4x - 5 = (x + 5)(x - 1) = 0$,
so the graph intersects the x-axis when $x = -5$ and $x = 1$.

Now you can use the graph to solve the inequality.
You're looking for $x^2 + 4x - 5 < 0$ — that's where the graph
is below the x-axis. You can see from the x-intercepts
that this is between the points $(-5, 0)$ and $(1, 0)$.
So the solution is **$-5 < x < 1$.**

b) **Find the values of x which satisfy both
$-x^2 - 2x + 4 > 2x - 1$ and $9x \leq 4x - 5$.**

Solve the second inequality in the normal way:
Subtract $4x$ from both sides: $9x - 4x \leq 4x - 5 - 4x$
$5x \leq -5$

Then divide both sides by 5: $5x \div 5 \leq -5 \div 5$
$x \leq -1$

To solve both inequalities, add the line
$x = -1$ to the sketch of $y = x^2 + 4x - 5$.

You're looking for where $x^2 + 4x - 5 < 0$ **and** $x \leq -1$
— so where the curve is below the x-axis,
and to the left of $x = -1$.

So the set of values is **$-5 < x \leq -1$.**

There's more on showing
inequalities using graphs on
the next couple of pages.

PRACTICE QUESTIONS

I'm no Picasso — but I can sketch a mean quadratic...

1) Solve the following linear inequalities:
   a) $4 - 8x \leq -10x - 6$ b) $-13 \leq 2x - 3 < 11$ c) $-11 < 1 - 3x < 7$

2) Use a number line to find the values of x which satisfy both $2x - 4 < 3x$ and $5 \geq 7x - 2$.

3) Solve the following quadratic inequalities:
   a) $3x^2 - 8 < 4$ b) $2x - x^2 + 8 > 0$ c) $6x \geq 2x^2 + 4$ d) $-5x^2 + 4x + 2 < 6x - x^2$

4) Find the values of x which satisfy both $x^2 + 3x + 7 < 3 - 2x$ and $2(3 - 2x) > 14$.

Graphical Inequalities

Follow These **Four Easy Steps** to Show **Inequalities** on a **Graph**

1) Write each inequality as an **equation**.

   Replace the **inequality sign** with an **= sign** and write in the form "y = ..." if you need to.

2) Draw the **graph** of each equation.

   Use a **dotted line** if the inequality had a < or > sign and a **solid line** if it had a ≤ or ≥ sign.

3) Decide **which side** of each line is the correct one.

   Put the coordinates of a point into each inequality.
   If the inequality is **true**, the point is on the **correct side** of the line.

   Use the origin if possible.

4) **Shade** or **label** the correct region.

   Check carefully what you're asked to do in the question.

EXAMPLE

Draw and shade the region that satisfies all three inequalities below.

$$3x < 6 + y \qquad -2 + y \leq x \qquad y \geq -x$$

1) Write each inequality as an equation in the form 'y =':

   $$y = 3x - 6 \qquad y = x + 2 \qquad y = -x$$

2) Then draw the graphs. They're all linear equations, so they're just straight lines:

3) Decide which side of each line is correct by plugging some coordinates into each inequality. You can use the origin for $3x < 6 + y$ and $-2 + y \leq x$, but the origin is on the line $y = -x$ so use (1, 2) for $y \geq -x$:

   $3x < 6 + y \Rightarrow 0 < 6$ which is true.
   So the origin is on the correct side of the line.

   $-2 + y \leq x \Rightarrow -2 \leq 0$ which is true.
   So the origin is on the correct side of the line.

   $y \geq -x \Rightarrow 2 \geq -1$ which is true.
   So (1, 2) is on the correct side of the line.

4) Shade the region below $y = x + 2$, and above $y = 3x - 6$ and $y = -x$.

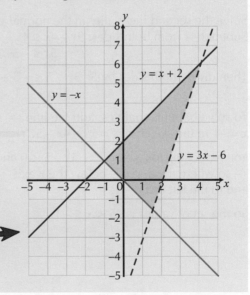

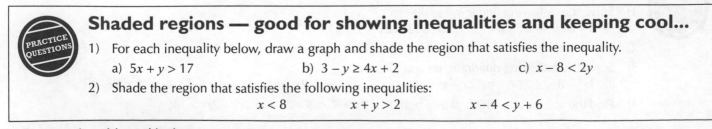

Shaded regions — good for showing inequalities and keeping cool...

PRACTICE QUESTIONS

1) For each inequality below, draw a graph and shade the region that satisfies the inequality.

   a) $5x + y > 17$ b) $3 - y \geq 4x + 2$ c) $x - 8 < 2y$

2) Shade the region that satisfies the following inequalities:

   $$x < 8 \qquad x + y > 2 \qquad x - 4 < y + 6$$

Graphical Inequalities

You can Show **Quadratic Inequalities** on a Graph   NEW CONTENT

At A-Level, you'll need to show **quadratic inequalities** on a graph as well as linear ones. To do this, you'll have to accurately sketch a **quadratic graph** using its x- and y-intercepts and turning point. Have a look at p.37 for how — the examples below will just tell you the key points.

EXAMPLES

Draw and label the region that satisfies the inequalities $y > x^2 - 4$ and $3x + 1 \geq y$.

Write the inequalities as equations in terms of y:

$$y = x^2 - 4 \qquad\qquad y = 3x + 1$$

Then draw the graphs. The quadratic graph is u-shaped. It intersects the y-axis at $y = -4$, and the x-axis at $x = 2$ and $x = -2$. The turning point is at $(0, -4)$.

Plug the origin $(0, 0)$ into each inequality:

$y > x^2 - 4 \Rightarrow 0 > -4$ which is true.
So the origin is on the correct side of the line.

$3x + 1 \geq y \Rightarrow 1 \geq 0$ which is true.
So the origin is on the correct side of the line.

Label the region above $y = x^2 - 4$ and below $y = 3x + 1$.

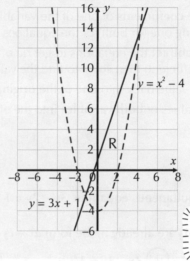

Make sure you get the dotted and solid lines right.

Draw and shade the region satisfying the following inequalities:

$$-x^2 + 3x - 2 > y \qquad\qquad x^2 < y + 9$$

Write each inequality as an equation in the form '$y =$':

$$y = -x^2 + 3x - 2 \qquad\qquad y = x^2 - 9$$

Then draw the graphs. $y = -x^2 + 3x - 2$ is n-shaped. It intersects the y-axis at $y = -2$ and the x-axis at $x = 1$ and $x = 2$. The turning point is at $(1.5, 0.25)$.

$y = x^2 - 9$ is u-shaped. It intersects the y-axis at $y = -9$ and the x-axis at $x = 3$ and $x = -3$. The turning point is at $(0, -9)$.

Find the region by substituting $(0, 0)$ into each inequality.

$-x^2 + 3x - 2 > y \Rightarrow -2 > 0$ which is false.
So the origin is on the wrong side of the line.

$x^2 < y + 9 \Rightarrow 0 < 9$ which is true.
So the origin is on the correct side of the line.

Shade the region below $y = -x^2 + 3x - 2$ and above $y = x^2 - 9$.

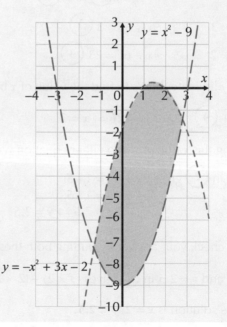

PRACTICE QUESTIONS

Or cook the inequalities a roast dinner — that'll satisfy 'em...

1) Draw and shade the regions that satisfy the following inequalities:
   a) $y > x^2$ and $8x + 4 < 2y$
   b) $-x^2 + 4 \geq y$ and $y + 10 < 14 - x$
   c) $8 - 2x^2 - y < 0$ and $6x > 3y - 18$
   d) $x^2 + 4y > 0$ and $x^2 - 6 < y - 2$

2) Draw and shade the region that satisfies the following inequalities:
   $$y < 3 + 5x - 2x^2 \qquad\qquad y \geq x^2 - x - 2$$

Simultaneous Equations

Use **Elimination** when the Simultaneous Equations are **Linear**

1) To solve a pair of **simultaneous equations**, you need to find the values of the **variables** (e.g. x and y) which will make **both** equations **true**.

2) For example, the simultaneous equations $x + 2y = 5$ and $3x - y = 1$ have the solution **$x = 1$ and $y = 2$** because plugging these values into the LHS of the equations gives the RHS.

Here's the **method** for solving simultaneous equations if both equations are **linear**:

1) Rearrange the equations into the form $ax + by = c$.

2) Make the **coefficients** of one of the variables the same by **multiplying** one (or both) of the equations by something.

In step 2, you can ignore the signs of the coefficients. E.g. you could make them 4 and −4.

3) **Add** or **subtract** the equations to get rid of the terms with the same coefficient, then **solve** to find the value of the **remaining variable**.

4) Plug this value back into one of the **original equations** to find the **other variable**.

5) Put the values of **both variables** into the **other original equation** to check your answer.

EXAMPLE

Solve the simultaneous equations $3x - 2y = 1$ and $2x + 3y = 11.5$.

Both equations are already in the form $ax + by = c$. Label them ① and ②.

$$①\ 3x - 2y = 1 \qquad\qquad ②\ 2x + 3y = 11.5$$

Make the coefficients of one of the variables the same.
Label the new equations ③ and ④:

$$①\times 3 \quad\rightarrow\quad 9x - 6y = 3 \;③$$
$$②\times 2 \quad\rightarrow\quad 4x + 6y = 23 \;④$$

Ling wished she could use elimination to solve all of her problems.

Now eliminate the y's to find the value of x by adding the two equations together:

$$③ + ④ \qquad 13x = 26 \Rightarrow x = 2$$

If the signs are the same, then subtract. If the signs are opposites, then add.

Putting this value of x into either equation will allow you to solve for y.

$$x = 2 \text{ in } ② \text{ gives } 2x + 3y = 11.5$$
$$4 + 3y = 11.5$$
$$3y = 7.5 \Rightarrow y = 2.5$$

Now check your answer by putting both these values into equation ①.

$$x = 2 \text{ and } y = 2.5 \text{ in } ① \text{ gives } (3 \times 2) - (2 \times 2.5) = 1 \;✔$$

So the solution is **$x = 2$, $y = 2.5$**.

PRACTICE QUESTIONS

Eliminating the x's — seems a tad cruel...

1) Solve the following simultaneous equations:

a) $y = 4x - 1$
$5x - 2y = 5$

b) $3x + 8y = 25$
$12x - 10y = 16$

c) $3x - 2y = 8$
$7x - 5y = 17$

d) $\frac{1}{2}x + 2y = 12$
$y = 4x - 11$

Simultaneous Equations

Use **Substitution** When One Equation is **Non-Linear**

If there are any squared terms in the equations, you can't use the elimination method — you have to use the **substitution** method instead. You might have seen this method at GCSE, but you'll use it at **A-Level** with trickier equations, e.g. with more than one squared variable.

1) Rearrange the **linear equation** to get a variable **on its own**.
2) **Substitute** this into the quadratic equation.
3) Rearrange into the **standard format** for a quadratic equation ($ax2 + bx + c = 0$) and solve (see p.16-18).
4) Plug these values back into the **linear equation** to find the values of the **other variable**.
5) Put the values of **both variables** into the **quadratic equation** to check your answer.

> At GCSE, you might have rearranged the quadratic equation instead. But you can't always get a variable on its own in the quadratic (see below), so I've said to rearrange the linear one here.

EXAMPLE

Solve the simultaneous equations $2y = 5 - x$ and $x^2 + y^2 = 10$.

Rewrite the linear equation so either x or y is by itself on one side of the equation.

Label the equations ① and ②.

$$① \quad x = 5 - 2y$$
$$② \quad x^2 + y^2 = 10$$

Substitute ① into ② to form another equation — call this ③.

$$x^2 + y^2 = 10$$
$$(5 - 2y)^2 + y^2 = 10 \quad ③$$

Then rearrange ③ to make a quadratic equation in the form $ax^2 + bx + c = 0$ and solve it.

$$25 - 10y - 10y + 4y^2 + y^2 = 10 \quad ③$$
$$5y^2 - 20y + 15 = 0$$
$$y^2 - 4y + 3 = 0$$
$$(y - 3)(y - 1) = 0, \text{ so } y = 3 \text{ or } y = 1$$

> You should factorise to solve the quadratic at this step if possible — you can use the quadratic formula if not.

To find the corresponding values of x, put each y-value back into the linear equation.

Substitute $y = 3$ into ① to give: $x = 5 - (2 \times 3) = 5 - 6 = -1$

Substitute $y = 1$ into ① to give: $x = 5 - (2 \times 1) = 5 - 2 = 3$

Check your answer by putting both pairs of values into equation ②.

$x = -1$ and $y = 3$ in ② gives $(-1)^2 + 3^2 = 10$ ✔

$x = 3$ and $y = 1$ in ② gives $3^2 + 1^2 = 10$ ✔

> There are two solutions, so the graphs of these equations will cross in two places.

So the solutions are **$x = -1$, $y = 3$ and $x = 3$, $y = 1$.**

Why did the chickens simultaneously cross the road*...

PRACTICE QUESTIONS

1) Solve the following simultaneous equations:
   a) $y = 4x + 4$ and $y = x^2 + 3x - 8$
   b) $y^2 - x^2 = 0$ and $3x - y = 3$
   c) $x^2 + xy - 10 = 0$ and $y + 2x = -7$

---

* [EDIT — don't worry, we've since had words with the 'Editor' responsible for that bad joke...]

Proof

Remember These **Important Facts** to Help You **Prove Things**

You came across proof at GCSE — but at A-Level you'll have to prove some **trickier** things.
So you need to make sure you're really comfortable with the **basic skills**.

These facts are a good starting point in a lot of proofs:

1) You can write any **even number** as $2x$.
2) You can write any **odd number** as $2x + 1$.
3) You can write **consecutive numbers** as x, $x + 1$, $x + 2$, etc.
4) You can show something is a **multiple** of a number, x, by showing it can be written '$x \times$ something'.
5) When you **add**, **subtract** or **multiply integers**, you'll always end up with an **integer**.

> x is just an integer (a whole number).

EXAMPLES

Prove that the difference between any two even numbers is always even.

Write the two even numbers as $2x$ and $2y$, where x and y are integers.

Now find the difference: $2x - 2y = 2(x - y)$. The difference between two integers is an integer, so $(x - y)$ is an integer.

$2x - 2y$ can be written as $2n$, where $n = x - y$. **This means that the difference between any two even numbers is a multiple of two, and is therefore even.**

Prove that, for any integer a, the expression $(3a - 1)(a + 2) + 2(a + 3)(a + 2)$ is always a multiple of 5.

Multiply out the brackets: $3a^2 + 5a - 2 + 2(a^2 + 5a + 6) = 3a^2 + 5a - 2 + 2a^2 + 10a + 12$

Combine like terms: $5a^2 + 15a + 10$

Take out a factor of 5: $5(a^2 + 3a + 2)$. As a is an integer, $a^2 + 3a + 2$ is also an integer.

The expression can be written as $5x$, where $x = (a^2 + 3a + 2)$, so it is a multiple of 5.

You can Use Facts to **Build Up** your **Argument** in a Proof

In some proofs, you'll need to use **facts** and **laws** from maths to show something is true.

> You'll learn more about this type of proof at A-Level — it's called 'proof by deduction'.

EXAMPLE

Prove that the result of dividing a rational number by any other rational number is also rational.

Call the two rational numbers x and y. Rational numbers can be written as fractions with integers on the top and bottom, so write $x = \frac{a}{b}$ and $y = \frac{c}{d}$ where a, b, c and d are integers, and b, c and d are not 0 (because you can't divide by 0).

$x \div y = \frac{a}{b} \div \frac{c}{d} = \frac{a}{b} \times \frac{d}{c} = \frac{ad}{bc}$. ad and bc are the products of two integers, so they are also integers. bc is not 0, as neither b nor c is zero.

The result of dividing two rational numbers can be written as a fraction with integers on the top and bottom and a non-zero denominator, so it is also rational.

Using proof by deduction, Ama worked out that her sister was the jumper-stealing culprit.

Proof

You can **Disprove** by Finding a **Counter Example**

Finding a **counter example** is one of the easiest ways to prove a statement is **wrong**. If you can just show **one case** of where the statement doesn't work, then you've disproved it. Sorted.

EXAMPLES

Find a counter example to disprove the following statement:
"When x is prime, $2x + 1$ is prime."

Try different primes in the expression.

Substituting $x = 3$ into $2x + 1$ gives $(2 \times 3) + 1 = 7$
7 is a prime number, so try again...

Substituting $x = 7$ into $2x + 1$ gives $(2 \times 7) + 1 = 15$
15 is not prime, so the statement is **not true**.

> It can sometimes take a while to find a counter-example, but just keep your cool...

By finding a counter example, disprove the following statement:
"Given that a and b are integers, if $a > b$, then $a^2 + 1 > b^2 + 1$."

At first glance, this might look correct — but take a closer look at the statement. a and b are integers, which means they can be negative numbers too...

So if $a = 1$ and $b = -2$, then $1 > -2$, so $a > b$.
But $a^2 + 1 = 2$ and $b^2 + 1 = 5$, so $a^2 + 1 < b^2 + 1$.

For $a = 1$ and $b = -2$, the second part of the statement isn't satisfied.

So this statement is **not true**.

Ian says, "If x and y are two different irrational numbers, then xy must also be irrational."
Find an example to show that Ian is wrong.

The square roots of any non-square number are irrational, so try $x = \sqrt{8}$ and $y = \sqrt{2}$:

$xy = \sqrt{8} \times \sqrt{2} = \sqrt{8 \times 2} = \sqrt{16} = 4$

4 is not an irrational number, so Ian is **wrong**.

> You saw on p.6 that an irrational number is one that can't be written exactly as a fraction.

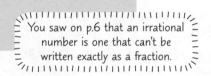

I looked for the proof in my pudding — but I just got really sticky...

1) Prove that $(n + 1)^2 - (n - 1)^2$ is even for any integer n.

2) Prove that the product of any two odd numbers is always odd.

3) Prove that the result of subtracting a rational number from any other rational number is also rational.

4) Phyllis believes the following statement is true: "$2^n + 1$ always gives a prime number, when n is an integer." Prove, by finding a counter example, that Phyllis is wrong.

5) Find a counter example to disprove the following statement: "For any pair of integers x and y, $\frac{x^2}{y^2} < \frac{x}{y}$."

Functions

Functions Take an **Input** and Give an **Output**

1) A **function** is a type of **mapping** — a set of **instructions** on how to get from **one number** to **another**. There are **two** main ways to write a function:

$$f(x) = 8x + 1$$
$$f : x \rightarrow 8x + 1$$

You might not have seen this notation at GCSE, so don't panic if you don't recognise it.

2) These both mean that 'f' takes a value for x, **multiplies** it by 8, **adds** 1, and **outputs** the result.

3) The set of values you put into a function is called the **domain**. The set of values output from a function is called the **range**. A function maps **each value** in the **domain** to **one** value in the **range**.

 NEW CONTENT

There's more about mappings, domains and ranges in the second year of A-Level Maths.

4) You can **evaluate** a function for different values by substituting each value into the function:

EXAMPLES

Find f(4) for the function $f(x) = (x + 6)^2$.

Just put $x = 4$ into the function and work out the answer: $f(4) = (4 + 6)^2 = 10^2 = \mathbf{100}$

$f : x \rightarrow 8x - 7$. **Find f(–3).**

Substitute $x = -3$ into the function:
$f(-3) = (8 \times -3) - 7 = \mathbf{-31}$

$f(x) = 2x^2 - 1$. **Find the value(s) of x when f(x) = 7.**

Set the function equal to 7 and solve:
$$2x^2 - 1 = 7$$
$$2x^2 = 8$$
$$x^2 = 4 \Rightarrow x = \mathbf{-2} \text{ or } x = \mathbf{2}$$

You can **Put Functions Together** to Make **Composite Functions**

1) When you have **two different functions**, say $f(x)$ and $g(x)$, you can combine them to make a **single new function**. This is called a **composite function**.

2) The composite function of $f(x)$ and $g(x)$ is written **fg(x)** — which means 'do **g first**, then **apply f** to your answer'. $ff(x)$ or $f^2(x)$ means 'do f **twice**'.

Composite functions were covered at GCSE, but there's some trickier stuff to learn about them in Year 2 of A-Level — so make sure you know the basics here.

3) You **always** do the function **closest to x** first.

EXAMPLES

$f(x) = x^2 + 20$ and $g(x) = 2x + 3$, find:
a) **fg(x)**
b) **gf(x)**

$fg(x) = f(g(x)) = f(2x + 3) = \mathbf{(2x + 3)^2 + 20}$
(or $4x^2 + 12x + 29$)

$gf(x) = g(f(x))$
$= g(x^2 + 20) = 2(x^2 + 20) + 3 = \mathbf{2x^2 + 43}$

Usually, you'll find that $fg(x) \neq gf(x)$.

$f(x) = x^2 + 1$, $g(x) = 4 - 2x$ and $h(x) = 3 + x$. **Find fgh(3).**

Start with the function closest to x and work outwards. So do h(3) first: $3 + 3 = 6$.
Then apply g to this result: $g(6) = 4 - (2 \times 6) = -8$
Then apply f to this result: $f(-8) = (-8)^2 + 1 = 65$, so $fgh(3) = \mathbf{65}$.

Functions

An **Inverse Function Reverses** the Effect of the **Original Function**

1) The **inverse** of a function **reverses** what the function does.
   The inverse of a function f(x) is written with the notation **f⁻¹(x)**.

   So if f(2) = 5, then f⁻¹(5) = 2.

2) You'll see inverse functions in more detail in Year 2 of A-Level.

3) You can **find the inverse** of a function using these steps:

   1) Write out the function 'f(x) = ...' but **replace** f(x) with x, and substitute y into the expression instead of x. This gives you $x = f(y)$.
   2) **Rearrange** to make y the **subject**.
   3) Then **replace** y with **f⁻¹(x)**.

EXAMPLE

If f(x) = 4x + 12, find f⁻¹(x).

Replace f(x) with x and x with y: $x = 4y + 12$

Then rearrange to make y the subject: $4y = x - 12 \Rightarrow y = \dfrac{x - 12}{4}$

Finally, replace y with f⁻¹(x) to get: **f⁻¹(x) = $\dfrac{x - 12}{4}$**

"Inverse <u>functions</u>?" cried Mark, "but those sound like the opposite of fun!"

If you **combine** a function f(x) with its **inverse** f⁻¹(x), you get **x**. $f^{-1}f(x) = ff^{-1}(x) = x$

This is because f(x) and f⁻¹(x) just **reverse** each other, so the **output** is the **original input**.
You can use this to check whether you've worked out an inverse function correctly.

EXAMPLE

a) If f(x) = 8 – 12x, find f⁻¹(2).

Find f⁻¹(x) — replace f(x) with x, and x with y:
$$x = 8 - 12y$$

Rearrange for y: $y = \dfrac{8 - x}{12}$

Replace y with f⁻¹(x) to get f⁻¹(x) = $\dfrac{8 - x}{12}$

Substitute $x = 2$ into the function: $\dfrac{8 - 2}{12} = \dfrac{1}{2}$

b) Show that ff⁻¹(x) = x for this function.

You know that f⁻¹(x) = $\dfrac{8 - x}{12}$,

so just put this into f(x) and simplify.

$$ff^{-1}(x) = f\left(\frac{8-x}{12}\right) = 8 - 12\left(\frac{8-x}{12}\right)$$
$$= 8 - (8 - x) = x$$

ff⁻¹(x) = x, so the inverse is correct. Hoorah.

These pages : words and numbers → expert function knowledge...

PRACTICE QUESTIONS

1) For the function g : $x \rightarrow \dfrac{11x + 13}{21 - 7x}$, evaluate g(2).

2) f(x) = 5x² + 12x – 7. Evaluate f(3).

3) If f(x) = 6x + 4 and g(x) = $\dfrac{28}{x}$, find:
   a) f(2)   b) g(4)   c) fg(x)   d) gf(x)

4) If f(x) = 4x² + 3 and g(x) = 6x – 6, find:
   a) fg(0)   b) gf(0)   c) f²(0)   d) g²(0)

5) Find the inverse functions of f(x) and g(x) below.
   a) f(x) = 9 – 11x   b) g(x) = $\dfrac{x - 4}{2}$

Straight Lines

The **Equation** of a **Straight Line** is **y = mx + c**

You need to know this stuff like the back of your hand for A-Level —
it comes up all over the place, from **logarithms** to **differentiation**.

If you know the **gradient** and the **y-intercept** of a straight line, you can give its **equation**.

> The equation of any straight line can be written in the form $y = mx + c$
> where m is the gradient of the line and c is the y-intercept of the line.

c is the **y-intercept** because it's the point on the graph where $x = 0$ — so $y = m \times 0 + c = c$.
Similarly, to find the x-intercept you set y equal to 0 and solve for x.

EXAMPLE

A straight line has a gradient of 4 and a y-intercept of 6.
Find the point where this line crosses the x-axis.

First find the equation of the line. The gradient $m = 4$ and y-intercept $c = 6$, so $y = 4x + 6$.
Now set y equal to zero and solve for x: $0 = 4x + 6 \Rightarrow 4x = -6 \Rightarrow x = -1.5$.
So the line crosses the x-axis at the point **(–1.5, 0)**.

You can also read off the gradient and y-intercept from the equation of a line. At **A-Level**,
you'll have to deal with straight-line equations written in **different forms**, e.g. $ax + by + c = 0$,
rather than $y = mx + c$. Just do a bit of rearranging to get the equation into the form you want.

EXAMPLE

Find the gradient of the line $3x + 4y + 4 = 0$.

Rearrange into $y = mx + c$ form: $4y = -3x - 4 \Rightarrow y = -\frac{3}{4}x - 1$. So the gradient is $-\frac{3}{4}$.

You can Find the **Gradient** using **Two Points** on a **Line**

You can take **any two points on a line** and use their coordinates to find the **gradient** of that line.

The gradient between two points
(x_1, y_1) and (x_2, y_2) is given by $\Longrightarrow$ $\dfrac{\text{change in } y}{\text{change in } x} = \dfrac{y_2 - y_1}{x_2 - x_1}$

EXAMPLE

Find the gradient of the straight line on the right.

Pick two accurate points on the line and label them
(x_1, y_1) and (x_2, y_2). Here, you can read off the points
$(x_1, y_1) = (-2, 3)$ and $(x_2, y_2) = (4, -2)$.

Plug these into the formula to find the gradient: $\dfrac{-2-3}{4-(-2)} = -\dfrac{5}{6}$

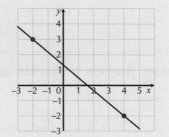

(🔘 PRACTICE QUESTIONS)

x goes to cross, but y intercepts — a strong defensive line...

1)   Find the gradient, x-intercept and y-intercept of the line $2x + y = -2$.
2)   A line has gradient 3 and y-intercept 5. Find the equation of the line.
3)   Find the gradient of the line that goes through the points (3, 4) and (–2, 1).

Straight Lines

Find the **Equation of a Line** using **Two Points** on the line

If you know **two points** that a line goes through, you can work out the **equation** of the line.

EXAMPLE

Find the equation of the line that passes through points (1, 7) and (3, 16).
Give your answer in the form $ax + by + c = 0$ where a, b and c are integers.

Label the points: $(x_1, y_1) = (1, 7)$ and $(x_2, y_2) = (3, 16)$

Find the gradient with $\dfrac{y_2 - y_1}{x_2 - x_1}$: $\dfrac{16 - 7}{3 - 1} = \dfrac{9}{2}$, so $y = \dfrac{9}{2}x + c$.

Substitute one of the points into the equation you've found and solve for c:

$y = \dfrac{9}{2}x + c \Rightarrow 7 = \dfrac{9}{2} + c \Rightarrow c = 7 - \dfrac{9}{2} = \dfrac{5}{2}$. This gives $y = \dfrac{9}{2}x + \dfrac{5}{2}$.

Multiply the whole equation by 2 to get integer coefficients: $2y = 9x + 5$.

Then rearrange into the correct form: **$9x - 2y + 5 = 0$**

Find the **Distance** between **Two Points** with **Pythagoras**

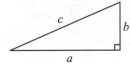

Pythagoras' theorem tells you that, for any **right-angled triangle**, $a^2 + b^2 = c^2$. So **length** $c = \sqrt{a^2 + b^2}$.

At A-Level you'll use this to calculate the magnitude of vectors...

You can use this to find the **distance** between **two points** (x_1, y_1) and (x_2, y_2):

1) Use the two points to make a **right-angled triangle**, with the **hypotenuse** as the line between the two points.

2) The **difference** between the **x-values** will give the length of one side. The **difference** between the **y-values** will give the length of the other side.

(x_2, y_2)

$\sqrt{(x_2 - x_1)^2 + (y_2 - y_1)^2}$

$y_2 - y_1$

(x_1, y_1) $x_2 - x_1$

Distance between (x_1, y_1) and (x_2, y_2) = $\sqrt{(x_2 - x_1)^2 + (y_2 - y_1)^2}$

EXAMPLE

Point A is (4, 17) and point B is (10, 2). What length is line AB?

Label the points: $(x_1, y_1) = (4, 17)$ and $(x_2, y_2) = (10, 2)$, so the length of AB

is given by $\sqrt{(10 - 4)^2 + (2 - 17)^2} = \sqrt{6^2 + (-15)^2} = \sqrt{261} = \textbf{16.16}$ (2 d.p.)

You can do a surprising amount with two points...

1) Find the equation of the line that passes through points (1, 2) and (5, 9).

2) A line is drawn between (–3, 4) and (7, –2). Find:
   a) The equation of this line in the form $ax + by + c = 0$ where a, b and c are integers.
   b) The length of the line. Give your answer to 2 d.p.

3) The distance Moe walks is modelled by a straight line graph.
   At time $t = 2$ hours, he has walked a distance of $d = 10$ kilometres.
   At time $t = 3.5$ hours, he has walked a distance of $d = 17.5$ kilometres.
   a) Give the equation of the line in the form $d = mt + c$.
   b) How long does it take Moe to travel 12.5 km?

Parallel and Perpendicular Lines

Parallel Lines have Equal Gradient

Parallel lines have the **same gradient**, which means they never cross each other.

At A-Level, you'll need these skills in parts of the differentiation topic, as well as in coordinate geometry.

EXAMPLE

> Line A has equation $y = 3x - 2$. Line B passes through point (4, 3) and is parallel to line A. Find the equation of line B.
>
> Parallel lines have the same gradient, so gradient of line B = gradient of line A = 3.
>
> This gives you the first bit of the equation for line B: $y = 3x + c$.
>
> You know the point (4, 3) is on the line, so plug this into the equation and solve for c:
> $3 = 3 \times 4 + c \Rightarrow 3 = 12 + c \Rightarrow c = 3 - 12 = -9$.
>
> So the equation of line B is $y = 3x - 9$.

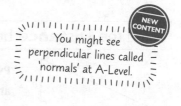

Chiara lived in a parallel universe where she couldn't be told off for crossing the line.

Perpendicular Lines Cross at Right Angles

NEW CONTENT

The **gradients** of perpendicular lines multiply together to give –1. So:

You might see perpendicular lines called 'normals' at A-Level.

> The gradient of the perpendicular line = –1 ÷ the gradient of the other line

EXAMPLE

> Find the equation of the line perpendicular to $3y - 2x + 4 = 0$ that passes through (2, 2). Give your answer in the form $ax + by + c = 0$, where a, b and c are integers.
>
> First, rearrange the equation into $y = mx + c$ form: $3y = 2x - 4$ so $y = \frac{2}{3}x - \frac{4}{3}$.
>
> The perpendicular line will have gradient $-1 \div \frac{2}{3} = -1 \times \frac{3}{2} = -\frac{3}{2}$.
>
> So the equation so far is: $y = -\frac{3}{2}x + c$.
>
> Plug in the point (2, 2) and solve for c: $2 = -\frac{3}{2} \times 2 + c$
>
> $\qquad\qquad\qquad 2 = -3 + c \Rightarrow c = 2 + 3 = 5$
>
> So the equation of the perpendicular line is $y = -x + 5$.
>
> Finally, rearrange into the form asked for in the question.
>
> Multiply by 2 to get rid of the fractions: $2y = -3x + 10$.
>
> Then put everything on one side of the equation: $3x + 2y - 10 = 0$.

Remember — when you divide by a fraction, you flip it over and multiply (see p.7).

PRACTICE QUESTIONS

Parallel lines meet about as often as the bad guy wins in Bond films...

1) Line F has equation $5y + 3x - 7 = 0$. Line G is parallel to line F and goes through the origin. What is the equation of line G?

2) Lines P and Q are perpendicular and meet at the point (–1, –2). The gradient of line P is 4. Find the equations of P and Q. Give your answers in the form $ax + by + c = 0$.

3) The points P (4, 3) and Q (9, 5) lie on line l_1. The line l_2 is perpendicular to l_1 and passes through Q. Find the equation of line l_2 in the form $ax + by + c = 0$.

Quadratic Graphs

All Quadratic Graphs have a Symmetrical Bucket Shape

You'll need to be really confident with **sketching quadratics** at A-Level — there's a whole topic on it.
You'll also use these skills in the **inequalities** and **kinematics** topics.

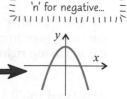
To remember, think 'n' for negative...

All **quadratic** graphs have the **same shape** — a **symmetrical curve**.

Positive quadratics (ones where the coefficient of x^2 is positive) are **u-shaped** like this. ➡

Negative quadratics (ones where the coefficient of x^2 is negative) are **n-shaped** like this. ➡

Work Out Three Things to Sketch A Quadratic

If you're asked to **sketch** a quadratic, you **don't** need to use graph paper and carefully plot every point. You just need to work out and label the **most important** bits of the graph to make sure your drawing is **roughly correct**. As well as the shape of the graph, you need to work out:

1) Where the graph **crosses** the **y-axis** (the **y-intercept**).

2) Where the graph **crosses** the **x-axis** (the **x-intercepts**). These are called the **roots** of the quadratic.

3) You might be asked to find the **turning point** (or '**vertex**') of the graph — this could be the **minimum** or **maximum** point depending on whether it's **u-shaped** or **n-shaped**.

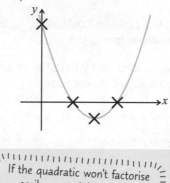

EXAMPLE

Sketch the graph of $y = x^2 - 5x + 6$, including any points of intersection with the axes and the vertex of the graph.

If the quadratic won't factorise easily, you might need to complete the square or use the quadratic formula (see p.18-21).

The coefficient of x^2 is positive, so the graph is **u-shaped**.

To find the y-intercept, let $x = 0$: $y = 0^2 - 5 \times 0 + 6 = 6$. So it **crosses the y-axis at $y = 6$**.

The graph intersects the x-axis when $y = 0$, so solve $x^2 - 5x + 6 = 0$ by factorising: $x^2 - 5x + 6 = (x - 2)(x - 3) = 0$, so the **x-intercepts** are $x = 2$ and $x = 3$.

Quadratic graphs are symmetrical, so the **x-coordinate** of the **turning point** of the graph is halfway between 2 and 3. So it's $(2 + 3) \div 2 = \textbf{2.5}$.

Plug this number back into the quadratic to find the **y-coordinate** of the **turning point**: $2.5^2 - 5 \times 2.5 + 6 = \textbf{-0.25}$. So the **turning point** is **(2.5, -0.25)**.

Now you can sketch the graph. Make sure to label each of the points.

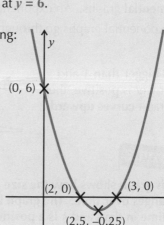
(0, 6)
(2, 0)   (3, 0)
(2.5, -0.25)

If a graph has a '**double**' root (i.e. factorises to the form $(x + a)^2$), this root is the **vertex**. The graph just **touches** the **x-axis** at this point — it doesn't cross it.

PRACTICE QUESTIONS

Symmetrical bucket hats — the must-have maths fashion trend...

1) Sketch these quadratics, labelling the points of intersection with the axes and the vertex of each graph.
    a) $y = x^2 - 3x$ b) $y = x^2 + x - 2$ c) $y = -x^2 + 6x - 9$ d) $y = 3x^2 + 2x - 8$

Harder Graphs

A **Cubic** Contains an **x^3** Term

Cubics are of the form $ax^3 + bx^2 + cx + d$ where a, b, c and d are all numbers ($a \neq 0$).
Cubic graphs have a characteristic '**wiggly**' shape.

If x^3 has a **positive** coefficient, then the graph goes from the **bottom left** to the **top right** of the axes.

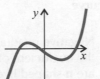

If x^3 has a **negative** coefficient, then the graph goes from the **top left** to the **bottom right**.

At **A-Level** you'll learn how to **factorise** and **sketch** simple cubics.
Here's an example of how to sketch one.

EXAMPLE

NEW CONTENT

Given that $x^3 + 3x^2 - x - 3 = (x + 1)(x - 1)(x + 3)$,
sketch the graph of $y = x^3 + 3x^2 - x - 3$.

The method for sketching a cubic is similar to the method for sketching a quadratic.

The x^3 term has a positive coefficient, so the graph will go **from the bottom left to the top right**.

Now you just need to find where the graph crosses the axes.
It **crosses the y-axis** when $x = 0$, so at $y = -3$. It crosses the x-axis when $y = 0$: $x^3 + 3x^2 - x - 3 = (x + 1)(x - 1)(x + 3) = 0$, so it **crosses the x-axis** at $x = -1$, **1** and -3.

Now you can use these values to sketch the graph.

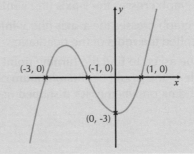

Exponential Graphs have Equation **$y = k^x$ or $y = k^{-x}$**

You'll cover lots on exponentials at A-Level.

Graphs of the form $y = k^x$ or $y = k^{-x}$ (for any positive number k) are called **exponential** graphs. An exponential graph is a **curve** that is always **above** the x-axis.
All exponential graphs go through the point **(0, 1)** — because anything to the power 0 is 1.

If k is **bigger than 1** and the power is **positive**, then the graph **curves upwards**.

If k is **between 0 and 1, OR** the **power is negative**, then the graph is **flipped horizontally**.

EXAMPLE

This graph shows how the size of a population of bacteria (**P**) changes over time. The graph has equation **P = kd**, where d is time in days and k is a positive constant. Find **k**.

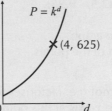

From the graph you can see that when $d = 4$, $P = 625$.
Substitute these values into the equation to find k:

$P = k^d \Rightarrow 625 = k^4 \Rightarrow k = \sqrt[4]{625} = \textbf{5}$

What goes up must come down — unless it's an exponential graph...

1) Sketch the graphs of the following cubics:
   a) $y = (x + 1)(x - 1)(x + 2)$ b) $y = (x - 1)(x - 2)(x - 3)$ c) $y = -x^3 - 5x^2 + 6x$

Harder Graphs

Reciprocal Graphs have Equation y = k/x or xy = k

1) Reciprocal graphs have **two curves** in **diagonally opposite quadrants**. They're **symmetrical** about the lines $y = x$ and $y = -x$.

2) The pair of quadrants the curves are in depends on whether k is **positive** or **negative**.

When k is **positive**, the graph looks like **this**.

When k is **negative**, the graph looks like this.

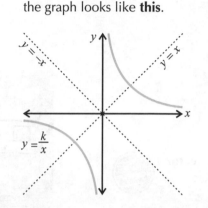

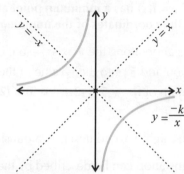

3) **Reciprocal graphs** are undefined at $x = 0$ (the y-axis) and $y = 0$ (the x-axis).

4) So reciprocal graphs **never touch** the axes — but they do get **infinitely close**. This means the axes are **asymptotes** of the graph — you'll be expected to know this term at A-Level. *[NEW CONTENT]*

Circles have Equations (x – a)² + (y – b)² = r² *[NEW CONTENT]*

The general equation for a **circle** with **centre** (a, b) and **radius** r is $(x - a)^2 + (y - b)^2 = r^2$.
Circle equations of the form $x^2 + y^2 = r^2$ were covered at GCSE.
These circles have centre $(0, 0)$, so are the **special case** where $a = b = 0$.

EXAMPLE

Find the equation of the tangent to $(x - 5)^2 + (y - 5)^2 = 50$ at the point $(-2, 4)$.

From the general equation of a circle, you know the centre is $(5, 5)$. Start by finding the gradient of the radius — the line from the centre of the circle to the point $(-2, 4)$.

Gradient of radius = $\dfrac{\text{change in } y}{\text{change in } x} = \dfrac{5-4}{5-(-2)} = \dfrac{1}{7}$.

A tangent to a circle meets a radius at 90°, so they are perpendicular (this is one of the circle theorems from GCSE).

This means the gradient of the tangent is $-1 \div \dfrac{1}{7} = -7$.

So the equation of the tangent so far is $y = -7x + c$.

Substitute in the point $(-2, 4)$ and solve for c:
$4 = -7 \times -2 + c \Rightarrow 4 = 14 + c \Rightarrow c = -10$.

So the equation of the tangent is $y = -7x - 10$.

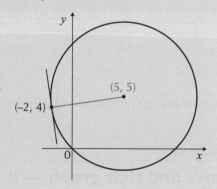

See page 36 for more on perpendicular lines.

Reciprocal graphs — a good way to get one over on your x...

PRACTICE QUESTIONS

1) Give the centre and radius of the following circles:
   a) $(x - 1)^2 + (y - 3)^2 = 16$
   b) $x^2 + (y + 2)^2 = 50$
   c) $(x + \tfrac{1}{2})^2 + (y - \tfrac{1}{2})^2 = 2$

2) Find the equation of the tangent to the circle $(x - 1)^2 + (y - 2)^2 = 8$ passing through point $(3, 4)$.

Graph Transformations

f(x) + a is a **Translation** Along the **y-axis**

A transformation of the form f(x) + a is a **translation parallel** to the **y-axis**.
The graph of f(x) moves **a units** in the y-direction (i.e. vertically),
so all of the **y-values** of f(x) will have **a added to them**.
Translating a graph doesn't change its shape, it just **moves** it.

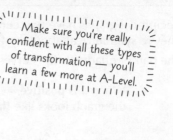
Make sure you're really confident with all these types of transformation — you'll learn a few more at A-Level.

EXAMPLE

The graph of y = f(x) has a minimum point at (2, –4).
Write down the coordinates of the minimum point of f(x) + 5.

This is a translation along the y-axis — so it only affects the y-values.

This means you add 5 onto the y-value of the minimum point.

So the minimum of f(x) + 5 is (2, –4 + 5) = **(2, 1)**.

At A-Level, you might have to describe a translation using a **column vector**. **NEW CONTENT**

This type of translation can be described by the vector $\begin{pmatrix} 0 \\ a \end{pmatrix}$.

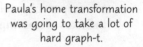
Paula's home transformation was going to take a lot of hard graph-t.

f(x + a) is a **Translation** Along the **x-axis**

A transformation of the form f(x + a) moves the graph of f(x)
left or **right** — i.e. **parallel** to the **x-axis**.

Be **careful** with these — when a is **positive**, the graph moves to the **left**.
When it is **negative**, the graph moves to the **right**.

EXAMPLE

The diagram shows the graph of y = f(x).
Sketch the graph of y = f(x + 3).

a = 3, so the graph moves 3 units in the negative x-direction.
This means the graph moves to the left by three units.

Even if you're given a graph that you don't recognise, just remember how the translations work and you'll be fine.

This type of translation can be described by the column vector $\begin{pmatrix} -a \\ 0 \end{pmatrix}$. **NEW CONTENT**

PRACTICE QUESTIONS

I can't find that graph — it must've been lost in translation...

1) The diagram on the right shows the graph of y = f(x):

   a) Sketch the following translations:

      i) f(x) + 3, labelling the coordinates of the minimum and where the curve meets the y-axis.

      ii) f(x – 2), labelling the coordinates of the minimum and where the curve meets the x-axis.

   b) State the column vectors that describe the above translations.

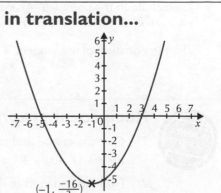
$(-1, \frac{-16}{3})$

Graph Transformations

–f(x) and f(–x) are both Reflections

$y = -f(x)$ is the **reflection** of $y = f(x)$ in the **x-axis**.
For **every point** on the graph of $y = f(x)$,
the **x-coordinate** stays the **same** and the
y-coordinate is **multiplied** by **–1**.

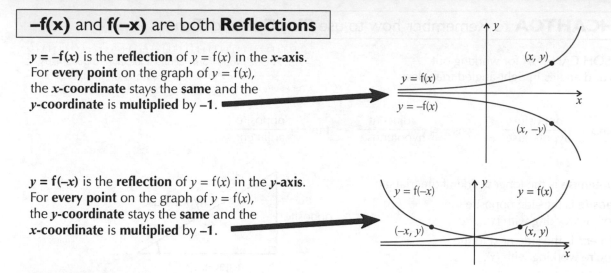

$y = f(-x)$ is the **reflection** of $y = f(x)$ in the **y-axis**.
For **every point** on the graph of $y = f(x)$,
the **y-coordinate** stays the **same** and the
x-coordinate is **multiplied** by **–1**.

EXAMPLES

a) $f(x) = x^2 + 2x - 8$.
Find the y-intercept of $y = -f(x)$.

The y-intercept of $f(x)$ is when $x = 0$.
$f(0) = (0)^2 + 2(0) - 8 = -8$, so is at $(0, -8)$.

The y-intercept of $y = -f(x)$ is the y-intercept
of $y = f(x)$ multiplied by –1.
So $y = -f(x)$ intersects the y-axis at **(0, 8)**.

b) Given that $f(x) = (x - 2)(x + 4)$,
find the roots of $f(-x)$.

The roots of $f(x)$ are $x = 2$ and $x = -4$.

$f(-x)$ is a reflection in the y-axis, so
all the x-values are multiplied by –1.
This means the roots of $f(-x)$ are
$x = -2$ and $x = 4$.

The diagram shows the graph of $y = g(x)$.
Sketch the graph of $y = -g(x)$.

The graph of $y = -g(x)$ is a reflection
of $y = g(x)$ in the x-axis.

The x-values stay the same, but the y-values
are multiplied by –1. This means the
y-intercept is 1 instead of –1.

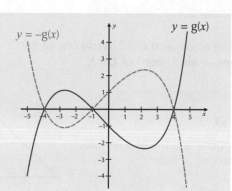

Now is a good time to stop and reflect on how far we've come...

1) The diagram on the right shows the graph of $y = f(x)$.
   Sketch the graphs of $y = -f(x)$ and $y = f(-x)$,
   labelling the x- and y-intercepts of each.

2) The quadratic graph $y = g(x)$ and its reflection $y = g(-x)$ are identical.
   For each of the following, state whether it could be $g(x)$.
   a) $x^2 - 4$ b) $x^2 + 3x - 1$ c) $-x^2$

3) $y = h(x)$ is the graph of a quadratic with the turning point $(4, -7)$.
   State the coordinates of the turning point of the following graphs:
   a) $y = -h(x)$ b) $y = h(-x)$

Trigonometry — Sin, Cos, Tan

Use **SOHCAHTOA** to Remember how to use **Sin, Cos** and **Tan**

Remember SOH CAH TOA for working out
side lengths and angles in right-angled triangles:

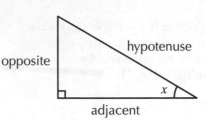

At A-Level, you'll come across a lot more trigonometry,
so you need to have a good grip on the basics.

$$\sin x = \frac{\text{opposite}}{\text{hypotenuse}} \qquad \cos x = \frac{\text{adjacent}}{\text{hypotenuse}} \qquad \tan x = \frac{\text{opposite}}{\text{adjacent}}$$

- The **hypotenuse** is the longest side of the triangle.
- The **opposite** is the side opposite the
  angle you're working with (x).
- The **adjacent** is the side next to the
  angle you're working with (x).

Which formula you need depends on the sides and angles you're given in the question.

EXAMPLE

Calculate the height of this isosceles triangle to 3 s.f.

Start by splitting the triangle down the middle
to get two right-angled triangles. Then you
can use trigonometry on one of them.

You have the adjacent side and want
to find the opposite side, so use tan:

$$\tan 55° = \frac{x}{5} \Rightarrow x = 5 \times \tan 55° = 7.1407... = \textbf{7.14 m (3 s.f.)}$$

To calculate a missing angle, you need to use one of the
inverse trig functions — **sin⁻¹, cos⁻¹** or **tan⁻¹**.

EXAMPLE

Find angle x to 3 s.f.

You know the opposite and the hypotenuse,
so you need to use sin:

$$\sin x = \frac{5}{7} \Rightarrow x = \sin^{-1}\left(\frac{5}{7}\right) = 45.5846... = \textbf{45.6° (3 s.f.)}$$

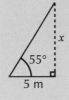

Maths rap group idea #23 — The Wu Tan(x) Clan...

PRACTICE QUESTIONS

1) Find the missing length l for each of these triangles. Give your answer to 3 s.f.

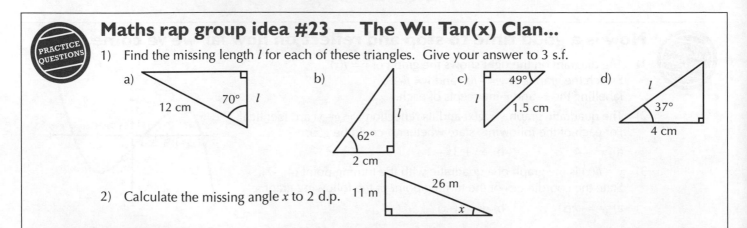

a)   12 cm   70° l

b) l 62°   2 cm

c)   49° l 1.5 cm

d) l 37°   4 cm

2) Calculate the missing angle x to 2 d.p.   11 m   26 m x

Trigonometry — Sin, Cos, Tan

Learn these Trig Values

There are **two triangles** that you can use to help you work out important trig values:

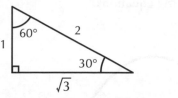

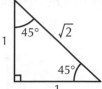

You can use Pythagoras' Theorem to check that the side lengths are right (p.35).

Using SOH CAH TOA on these triangles gives you these values:

$$\sin 30° = \frac{1}{2} \qquad \sin 60° = \frac{\sqrt{3}}{2} \qquad \sin 45° = \frac{1}{\sqrt{2}}$$

$$\cos 30° = \frac{\sqrt{3}}{2} \qquad \cos 60° = \frac{1}{2} \qquad \cos 45° = \frac{1}{\sqrt{2}}$$

$$\tan 30° = \frac{1}{\sqrt{3}} \qquad \tan 60° = \sqrt{3} \qquad \tan 45° = 1$$

Sarah found trigonometry difficult, so she decided to work on her tan.

You can **Calculate Trig Values** using the **Unit Circle**   NEW CONTENT

The **unit circle** is a circle with **radius 1** centred on the **origin**.
At A-Level, you'll see the unit circle used to calculate trig values.

1) You can take a **point** on the unit circle and make a **right-angled triangle**.

2) You know the **hypotenuse** of this triangle will be **1**, as it is the **radius** of the unit circle.

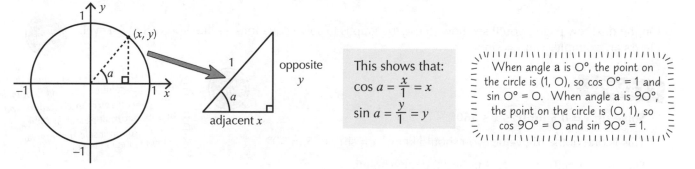

This shows that:
$$\cos a = \frac{x}{1} = x$$
$$\sin a = \frac{y}{1} = y$$

When angle a is 0°, the point on the circle is (1, 0), so cos 0° = 1 and sin 0° = 0. When angle a is 90°, the point on the circle is (0, 1), so cos 90° = 0 and sin 90° = 1.

3) This means you can find the **coordinates** of any point (*x, y*) on the unit circle by using trig — the coordinates are given by **(cos a, sin a)**.

4) The angle *a* needs to be measured **anticlockwise** from the **positive x-axis**.

EXAMPLE

**Find the coordinates of point P on the unit circle.
Give your answer to 2 d.p.**

The point is on the unit circle, so you know
the *x*-coordinate is cos 130° = –0.64278... and
the *y*-coordinate is sin 130° = 0.76604...)

So the coordinates of P to 2 d.p. are **(–0.64, 0.77)**

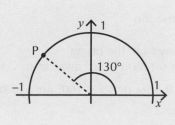

More circles and triangles than a PlayStation® cheat code...

1) The diagram on the right shows a point Q that lies on the unit circle.
   What are the coordinates of Q?
   Give your answer to 2 d.p.

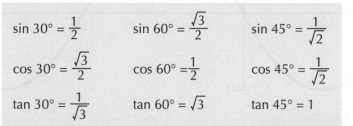

Trigonometry — Graphs

| The **Graph** of **y = sin x** Repeats Every **360°** |

You should have seen **all three** of the trig graphs at GCSE. You'll need to be really comfortable with them at A-Level, because they're used for **solving trig equations**.

This is the graph of $y = \sin x$:

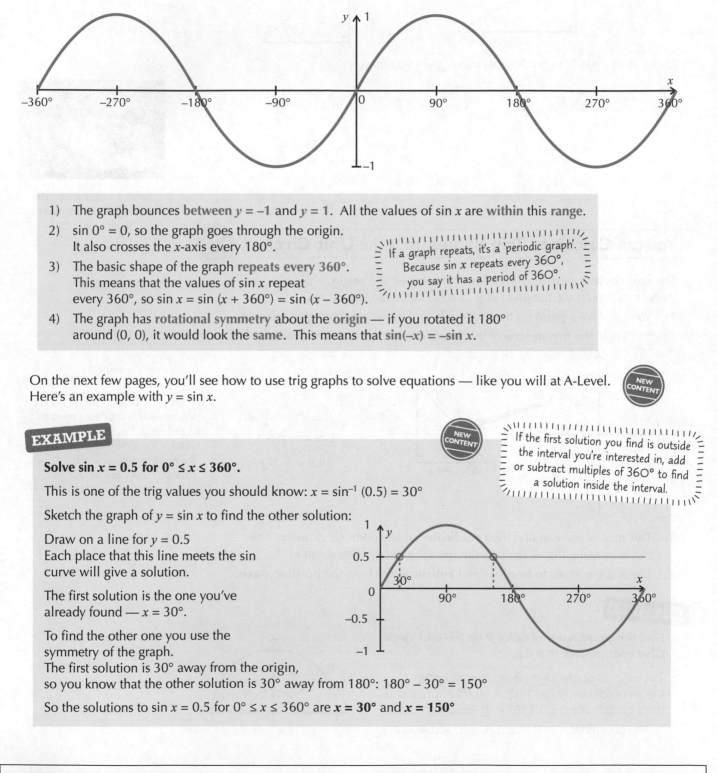

1) The graph bounces **between** $y = -1$ and $y = 1$. All the values of sin x are **within** this **range**.

2) sin 0° = 0, so the graph goes through the origin. It also crosses the x-axis every 180°.

3) The basic shape of the graph **repeats every 360°**. This means that the values of sin x repeat every 360°, so $\sin x = \sin(x + 360°) = \sin(x - 360°)$.

If a graph repeats, it's a 'periodic graph'. Because sin x repeats every 360°, you say it has a period of 360°.

4) The graph has **rotational symmetry** about the **origin** — if you rotated it 180° around (0, 0), it would look the **same**. This means that $\sin(-x) = -\sin x$.

On the next few pages, you'll see how to use trig graphs to solve equations — like you will at A-Level. Here's an example with $y = \sin x$.

NEW CONTENT

EXAMPLE

NEW CONTENT

If the first solution you find is outside the interval you're interested in, add or subtract multiples of 360° to find a solution inside the interval.

Solve sin x = 0.5 for 0° ≤ x ≤ 360°.

This is one of the trig values you should know: $x = \sin^{-1}(0.5) = 30°$

Sketch the graph of $y = \sin x$ to find the other solution:

Draw on a line for $y = 0.5$
Each place that this line meets the sin curve will give a solution.

The first solution is the one you've already found — $x = 30°$.

To find the other one you use the symmetry of the graph.
The first solution is 30° away from the origin, so you know that the other solution is 30° away from 180°: 180° − 30° = 150°

So the solutions to sin $x = 0.5$ for 0° ≤ x ≤ 360° are **x = 30°** and **x = 150°**

PRACTICE QUESTIONS

It'd be a sin not to learn every last bit of this page...

1) By sketching a graph, find all the solutions to the following equations in the interval 0° ≤ x ≤ 360°. Where necessary, give your answers to 1 decimal place.

   a) $\sin x = \dfrac{1}{\sqrt{2}}$ b) $\sin x = -0.7$ c) $\sin x = -0.2$

Trigonometry — Graphs

The **Graph** of **y = cos x** is the **Same Shape** as **sin x**

The graph of **cos x** is the **same shape** as the graph of **sin x**, but shifted left by 90°.
So the graph of $y = \cos x$ looks like this:

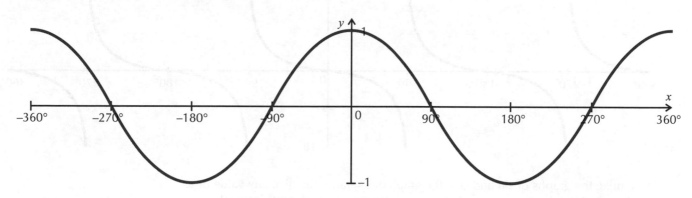

Because it's the same shape as the sin graph, it shares some of the same properties:

1) The *y*-values for the graph are all **between –1 and 1**.

2) The basic shape of the graph **repeats every 360°**, so $\cos x = \cos(x + 360°) = \cos(x - 360°)$.

But there are some differences between the two:

1) cos x is **symmetrical** about the *y*-axis — you can **reflect** it in the *y*-axis and it'll look the same. So $\cos(-x) = \cos(x)$.

2) Because cos 0° = 1, the graph crosses the *y*-axis at *y* = 1.

3) It **crosses the *x*-axis at ± 90°, ± 270°** etc.

NEW
CONTENT

Just like with sin x, if the first solution you find is outside the interval, add or subtract multiples of 360° to find a solution inside the interval.

EXAMPLE

Solve cos x = –0.8 for –180° ≤ x ≤ 180°. Give your answers to 1 d.p.

Solving cos x = –0.8 with a calculator will give you $x = \cos^{-1}(-0.8) = 143.1301...°$

This gives one answer. Now sketch the graph over this interval to find the second solution.

As every 360° interval of the graph is symmetrical, the second solution will be the same distance to the left of the *y*-axis as the first is to the right.
The first solution is 143.1301...° to the right of the *y*-axis, so the second solution is 143.1301...° to the left of the *y*-axis. So x = –143.1301...°

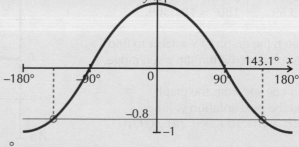

So the solutions to cos x = –0.8 for –180° ≤ x ≤ 180°
are **x = 143.1° and x = –143.1° (1 d.p.)**

PRACTICE QUESTIONS

Cosimodo — The humpback trig graph of Notre Dame...

1) By sketching a graph, find all the solutions to the following equations in the interval –360° ≤ x ≤ 360°. Where necessary, give your answers to 1 decimal place.

     a) $\cos x = \dfrac{1}{\sqrt{2}}$ b) $\cos x = 0.1$ c) $\cos x = -0.4$

Trigonometry — Graphs

The **Graph** of **y = tan x** Repeats Every **180°**

The graph of **y = tan x** is completely different from the other two. It looks like this:

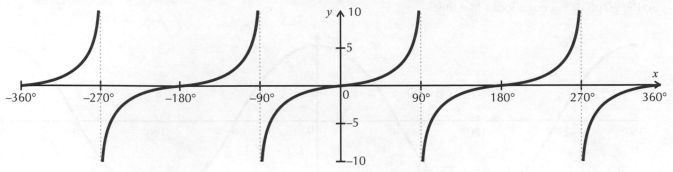

1) Unlike the graphs of sin and cos, the graph of $y = \tan x$ can give **any value** of y. The graph takes **every** y value between **−∞** and **+∞** in each **180° interval**.

2) The graph **repeats every 180°** and goes through the origin (because tan 0° = 0).

3) The graph is **not defined** at ± 90° and at ± 270°:

- These points are marked on the graph by dotted lines called **asymptotes**.
- As the graph approaches each asymptote from the **left**, it shoots **up to infinity**.
- As it approaches from the **right**, it shoots **down to minus infinity**.
- The graph **never** actually **touches** the asymptotes, although it does get **infinitely close**.

NEW CONTENT

EXAMPLE

NEW CONTENT

Solve tan x = −5 for 0° ≤ x ≤ 360°. Give your answer to 2 d.p.

Use your calculator to get a solution to the equation: $\tan^{-1}(-5) = -78.69...$
This is outside the interval, so
add 180° to find the first solution:
$-78.69...° + 180° = 101.309...°$

Sketch the graph of $y = \tan x$ to find
the solutions within the given range.

You can see from the graph
that the next solution will be
$180° + 101.309...° = 281.309...°$

So the solutions are $x = 101.31°$ and $x = 281.31°$ (2 d.p)

Fake tan x — it's not quite orange, but it gets infinitely close...

PRACTICE QUESTIONS

1) By sketching a graph, find all the solutions to the following equations in the interval −360° ≤ x ≤ 360. Where necessary, give your answers to 1 decimal place.

   a) tan x = −1   b) tan x = 7   c) tan x = −6

The Sine and Cosine Rules

The **Sine** and **Cosine Rules** work for **Any Triangle**

1) To work out side lengths and angles in **any triangle** (not just right-angled ones), you'll need to use the **sine** and **cosine rules**. These came up at GCSE, but you need to be slick at using them for A-Level.

2) Be **careful** how you **label** the triangle when using the sine and cosine rules — the side **opposite** an angle should have the **same letter**. E.g. side *b* is opposite angle *B*.

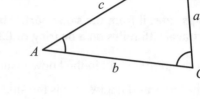

The Sine Rule
$\dfrac{a}{\sin A} = \dfrac{b}{\sin B} = \dfrac{c}{\sin C}$

The Cosine Rule
$a^2 = b^2 + c^2 - 2bc \cos A$

Which **rule** you use depends on which **sides** and **angles** you know in the **triangle**.

Learn when to use the **Sine Rule**

If you have **two angles** and a **side**, you use the **sine rule** to find the lengths of the other sides. Because you have two angles, you can find the **third** by **subtracting** them from **180°**.

EXAMPLE

Find the length *AC* in the triangle shown.

First, find the third angle in the triangle:
$180° - 42° - 66° = 72°$.

By the sine rule: $\dfrac{12}{\sin 72°} = \dfrac{b}{\sin 42°}$.

Rearrange to give: $b = \dfrac{12 \times \sin 42°}{\sin 72°} = $ **8.4 cm (1 d.p.)**

When Sam saw the cloud, he knew it must be a sine.

If you're given **two sides** and an **angle** that's **not between them**, you need to use the **sine rule**.

EXAMPLE

Find angle *BAC* in the triangle shown.

The known angle is not enclosed by the two known sides, so use the sine rule. Find angle *C* first:

$$\dfrac{b}{\sin B} = \dfrac{c}{\sin C} \Rightarrow \dfrac{11}{\sin 41°} = \dfrac{7}{\sin C}$$

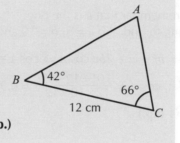

Rearrange to get $\sin C = \dfrac{7 \times \sin 41°}{11} = 0.417... \Rightarrow C = \sin^{-1}(0.417...) = 24.676...°$

This means that angle $BAC = 180° - 41° - 24.676...° = $ **114.3° (1 d.p.)**

All these rules man — they're bringing me down...

PRACTICE QUESTIONS

1) For each of the following triangles, find the missing value *x* to 1 decimal place:

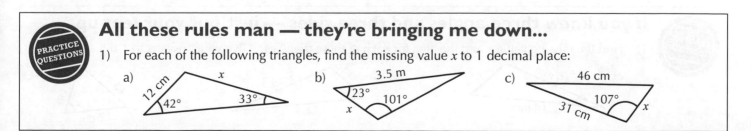

a) 12 cm, *x*, 42°, 33°
b) 3.5 m, 23°, 101°, *x*
c) 46 cm, 107°, 31 cm, *x*

The Sine and Cosine Rules

Learn when to use the Cosine Rule

If you're given **two sides** and the **angle between them**,
you can use the **cosine rule** to find the length of the third side.

EXAMPLE

**Two ships set sail from the same port. Ship X travels 42 miles due north.
Ship Y travels 36 miles on a bearing of 035°. How far apart are the two ships?**

Draw a diagram and fill in the known values.

Label the sides and angles — z is the side you're trying to find.

From the diagram, you can see that the two known lengths are
either side of the known angle, so you need to use the cosine rule.

$z^2 = 36^2 + 42^2 - 2 \times 36 \times 42 \times \cos 35° = 582.884....$

So $z = \sqrt{582.884...} = $ **24.1 miles (1 d.p.)**

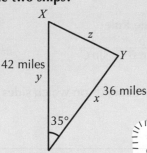

Bearings are measured clockwise from the north line (a vertical line).

If you know **all three sides** of the triangle but no angles, then you need the **cosine rule**.

EXAMPLE

Find angle A in the triangle shown.

Label the sides a, b and c — remember that a is the one
opposite the angle you're trying to find. So $a = 5$, $b = 10$ and $c = 12$.

Rearrange the cosine rule: $a^2 = b^2 + c^2 - 2bc \cos A \Rightarrow \cos A = \dfrac{b^2 + c^2 - a^2}{2bc}$

Plug these into the cosine rule: $\cos A = \dfrac{10^2 + 12^2 - 5^2}{2 \times 10 \times 12} = 0.9125$

So $A = \cos^{-1} 0.9125 = $ **24.1° (1 d.p.)**

There's also a Formula for the Area of a Triangle

The **area** of a triangle is given by this formula:

To use the **formula**, you need **one angle** and the **two sides
enclosing the angle**. If you don't have all of these,
you'll need to use the sine or cosine rule to find them.

$$\text{Area} = \frac{1}{2}ab \sin C$$

EXAMPLE

Find the area of this triangle to 1 d.p.

You're given an angle and two sides enclosing it,
so you can plug the numbers straight into the formula.

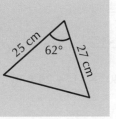

So the area is $\frac{1}{2} \times 25 \times 27 \times \sin 62° = $ **298.0 cm² (1 d.p.)**

PRACTICE QUESTIONS

If you know three angles and three sides — just put your feet up...

1) Find the missing value x and the area for each of these triangles. Give your answers to 1 d.p.

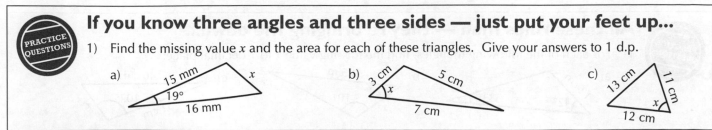

a) 15 mm, 19°, 16 mm, x

b) 3 cm, 5 cm, x, 7 cm

c) 13 cm, 11 cm, x, 12 cm

Vectors

Vectors have a Direction and a Magnitude

You will have met **vectors** at GCSE, but you're going to see a whole lot more of them at A-Level, especially in **mechanics**. They're often used for modelling things like forces.

1) **Scalars** are quantities without a direction — e.g. a **speed** of 5 m/s. They're just **numbers**.

2) **Vectors** represent a movement of a certain **magnitude (size)** in a **direction** — e.g. if two objects have a **velocity** of 5 m/s and –5 m/s, this means they are travelling at the **same speed** but in **opposite directions**.

There are Different Ways to Represent Vectors

1) Vectors are **drawn** as **arrows**, like this: 3 m/s →
   - The **direction** of the vector is shown by the **arrowhead**.
   - The **size** of the vector is shown by the line's **length**.

At GCSE you may have also seen vectors written like this — a̰.

2) Vectors are **written** as a lowercase **bold** letter (**a**) or a lower case **underlined** letter (a̲).

3) If the **start** and **end point** of a vector are known, it might also be written like this: $\overrightarrow{AB}$.

4) Vectors can also be written as column vectors — $\overrightarrow{AB}$ $\overrightarrow{BA}$

   e.g. $\begin{pmatrix} 4 \\ 1 \end{pmatrix}$ is a vector that goes **four units right** and **one unit up**.

Adding Vectors Describes Movements Between Points

1) Adding two vectors **a** + **b** means 'go along **a** then go along **b**'.

2) To add two vectors you can draw their arrows **nose to tail**.

3) The **single** vector that goes from the **start** of the first vector in the sum to the **end** of **the last** is called the **resultant vector**.

4) **Subtracting** one vector from another is a little more complicated — because –**b** is just the vector **b reversed**, you can think of **a** – **b** as **a** + (–**b**). So **a** – **b** means 'go along **a** then backwards along **b**'.

EXAMPLE

Find $\overrightarrow{AD}$ in terms of s, t and u.

The first thing to do is relabel the vectors so they're nose to tail from A to D like this.

Drawing a diagram really helps with vector questions.

Now you can see that to get from A to D, you go along –**s**, **t** and then –**u**:

$\overrightarrow{AD} = -s + t + (-u) = t - s - u$

The order you write the vectors in doesn't matter.

Woe to the scalars — and to the vector go the spoils...

1) $\overrightarrow{AB} = 2s + t$ and $\overrightarrow{BC} = 2t - \frac{1}{2}s$.
   Show that $2\overrightarrow{AC} = 3(s + 2t)$.

2) Using the diagram on the right, find $\overrightarrow{TW}$ in terms of the vectors **a** and **b**.

Vectors

Multiplying a Vector by a Scalar Changes its Size

1) Multiplying a vector by a **positive scalar** changes the vector's **size**, but **not** its direction.

2) Multiplying a vector by a **negative scalar** changes the vector's **size** and its **direction** gets **switched**.

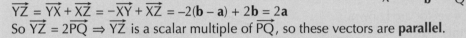

3) To multiply a **column vector** by a scalar, multiply the top and bottom numbers **separately** by the scalar, e.g. $2 \times \begin{pmatrix} 4 \\ 1 \end{pmatrix} = \begin{pmatrix} 2 \times 4 \\ 2 \times 1 \end{pmatrix} = \begin{pmatrix} 8 \\ 2 \end{pmatrix}$.

4) To show **two vectors** are **parallel**, you need to show they are **scalar multiples** of each other.

EXAMPLE

$\overrightarrow{XQ} = \mathbf{b}$ and $\overrightarrow{PQ} = \mathbf{a}$. Point P lies halfway along $\overrightarrow{XY}$ and point Q lies halfway along $\overrightarrow{XZ}$. Show that $\overrightarrow{YZ}$ is parallel to $\overrightarrow{PQ}$.

As Q is the midpoint of $\overrightarrow{XZ}$, $\overrightarrow{XZ} = 2\overrightarrow{XQ} = 2\mathbf{b}$.
$\overrightarrow{XP} = \mathbf{b} - \mathbf{a}$. P is the midpoint of $\overrightarrow{XY}$, so $\overrightarrow{XY} = 2\overrightarrow{XP} = 2(\mathbf{b} - \mathbf{a})$.

$\overrightarrow{YZ} = \overrightarrow{YX} + \overrightarrow{XZ} = -\overrightarrow{XY} + \overrightarrow{XZ} = -2(\mathbf{b} - \mathbf{a}) + 2\mathbf{b} = 2\mathbf{a}$
So $\overrightarrow{YZ} = 2\overrightarrow{PQ} \Rightarrow \overrightarrow{YZ}$ is a scalar multiple of $\overrightarrow{PQ}$, so these vectors are **parallel**.

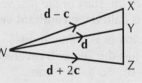

You can use Vectors to Show Three Points are on a Line

1) If **three or more points** all lie on a **single straight line**, they are **collinear**.

2) To show three points A, B and C are collinear, you need to show that $\overrightarrow{AB}$ and $\overrightarrow{BC}$ are parallel.

3) So you need to show that the vectors are **scalar multiples** of one another.

EXAMPLE

Show that X, Y and Z are collinear.

First, find $\overrightarrow{XY}$ and $\overrightarrow{YZ}$:

$\overrightarrow{XY} = -(\mathbf{d} - \mathbf{c}) + \mathbf{d} = \mathbf{c}$ and $\overrightarrow{YZ} = -\mathbf{d} + \mathbf{d} + 2\mathbf{c} = 2\mathbf{c}$.
$\overrightarrow{YZ} = 2\overrightarrow{XY}$, which means $\overrightarrow{XY}$ and $\overrightarrow{YZ}$ are parallel, so X, Y and Z are collinear.

A Vector's Magnitude is its Length

NEW CONTENT

> The magnitude of a vector **a** is written $|\mathbf{a}|$. The magnitude of $\overrightarrow{AB}$ is written $|\overrightarrow{AB}|$.

At **A-Level** you'll have to calculate the **magnitude of a vector**.
To do this, you need to work out the **distance** between the **start point** and the **end point**.
You do this using Pythagoras' theorem (see page 35).

Magnitude of vector $\begin{pmatrix} a \\ b \end{pmatrix} = \sqrt{a^2 + b^2}$

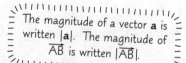

Getting from point A to point C has never been so exciting...

PRACTICE QUESTIONS

1) Show that the vectors $\mathbf{v} = 4\mathbf{a} + 6\mathbf{b}$ and $\mathbf{u} = 6\mathbf{a} + 9\mathbf{b}$ are parallel.

2) Using the diagram on the right, show that W, X and Y are collinear.

3) Calculate the magnitude of the following vectors: a) $\begin{pmatrix} 3 \\ 2 \end{pmatrix}$ b) $\begin{pmatrix} -1 \\ -1 \end{pmatrix}$

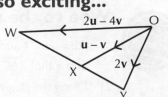

Sampling

A **Sample** Tells You **About a Population**

In A-Level Maths, you'll cover sampling in **more detail** and learn some **new sampling methods**. Here's a **recap** on what sampling is, with a couple of new terms thrown in.

1) The **population** is the **whole group** you want to investigate.
2) If you collect information from **every single member** of the population, it's called a **census**. A census gives really **accurate** results, but it's usually **too difficult** or **time-consuming** to survey everyone.
3) That's where **sampling** comes in — you can choose a **selection** from the population to **represent** the whole group. This smaller group is called a **sample**.
4) First you need to identify all the people or things you can **sample from** — these are the **sampling units**, and a **full list** of the sampling units is called the **sampling frame**. For reliable results, your sampling frame should include as much of the population as possible.

You have to **Choose** your Sample **Carefully**

1) It's really **important** that a sample is **similar** to the population. This means that the sample is **representative** of the population, which allows you to use it to **draw conclusions** about the whole **population**.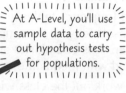

At A-Level, you'll use sample data to carry out hypothesis tests for populations.

2) A **biased** sample is one that **doesn't fairly represent** the population. It can be hard to avoid bias altogether, but there are a few **rules** you can follow to **limit it** as much as possible.

- Choose from the **correct population** and don't **exclude** anyone.
- Choose your sample **at random**.
- Make sure your sample is **big enough**.

3) You need to be able to **spot sampling bias** — think about **when**, **where** and **how** the sample is taken, and if it's **big enough**.

Choose a **Random** Sample using **Simple Random Sampling**

One way to get a **random** sample is to use the **simple random sampling** method. This gives every member of the population an **equal chance** of being in the sample, so it's **unbiased**.

To **select** a simple random sample, start by giving **every** member of the population a **unique number**. Then **generate a list of random numbers** and **match** them to the population members.

EXAMPLE

A retailer has 700 store card holders. Describe how simple random sampling could be used to select a sample of 30 of the card holders to survey.

Start by drawing up a list of all the card holders, and give each one a 3-digit number between 001 and 700.

Next, generate a list of 3-digit random numbers, using a calculator or random number table. You should reject any numbers over 700, or any that are repeated.

Stop when you have a list of 30 unique numbers, and select the card holders with the matching numbers — this is your sample.

Vicky decided to eliminate bias by sampling all of the tarts.

Choose sample members randomly — it's common census...

1) A council wants to survey the residents of a town about local issues. They hold a meeting one lunchtime where residents can give their views. If the council uses the residents who attend the meeting as their sample, explain why this sample may be biased.

Data Basics

Learn the **Difference** between **Quantitative** and **Qualitative** Data

1) A **variable** is a quantity that can take **different** values. There are two broad **categories**:

    Qualitative variables take non-numerical values — e.g. colours of cars.
    Quantitative variables take numerical values — e.g. temperature.

Get familiar with the basics — you'll need them for the A-Level statistics topics.

2) There are two different types of **quantitative data** — **discrete** and **continuous**:

    Discrete variables take certain values within a particular range — e.g. shoe size.
    Continuous variables can take any value within a particular range — e.g. length.

You need to Know about **Two Types** of **Frequency Table**

1) **Frequency tables** show how many times a data value occurs.
2) **Grouped frequency tables** group data into **classes**.
   These classes tell you the **range** that the value is contained in.
   The **upper class boundaries** and **lower class boundaries** can
   be defined with inequalities to cover all possible values.

Mass (m, g)	Frequency
$150 \leq m < 180$	2
$180 \leq m < 250$	6
$250 \leq m < 300$	12

lower class boundary   *upper class boundary*

3) You can work out the **class width** with this formula:

    Class width = upper class boundary – lower class boundary

4) You can also calculate the **mid-point of a class**.

$$\text{Mid-point} = \frac{\text{upper class boundary} + \text{lower class boundary}}{2}$$

EXAMPLE

The masses of 21 rocks (to the nearest 10 g) are recorded in the table. Add columns to the table to show the upper and lower class boundaries, class widths and the class mid-points.

Mass (g)	Frequency
150-250	7
260-320	9
330-400	5

The lightest rock that would weigh 150 g to the nearest 10 g is 145 g.
So the lower class boundary of the 150-250 class is 145 g.

The upper class boundary of the 150-250 class is 255 g.
This is the same as the lower class boundary of the
260-320 class, so there are no gaps between classes.

The data has been rounded, so the table doesn't show the actual class boundaries.

Find the class boundaries for the other classes, then use the
formulas to calculate each class width and mid-point:

These classes can be written using inequalities. E.g. the 150-250 class would be written $145 \leq m < 255$.

Mass (g)	Frequency	Lower class b'dary (g)	Upper class b'dary (g)	Class width (g)	Mid-point (g)
150-250	7	**145**	**255**	**110**	**200**
260-320	9	**255**	**325**	**70**	**290**
330-400	5	**325**	**405**	**80**	**365**

What kind of quantitative data do spies use? Discreet variables...

1) The table on the right shows the length
   (to the nearest 10 mm) of 50 pieces of string.
   Add four columns to show the lower class
   boundaries, upper class boundaries, the
   class widths and the class mid-points.

Length (mm)	Frequency
80-110	15
120-200	16
210-320	19

Histograms

Histograms Show Frequency Density

You'll have seen histograms at GCSE — but they rear their ugly heads again at A-Level.

1) **Histograms** are like bar charts, but the **height** of each bar tells you the **frequency density** instead of the frequency. You can calculate frequency density with this **formula**:

> Frequency Density = Frequency ÷ Class Width

Histograms are for continuous data.

2) The **width** of the bars tells you the **class widths**.

3) The **area** of each of the bars tells you the **frequency**. This is because you can **rearrange** the **formula** for frequency density to give: **Frequency = Frequency Density × Class Width**.

EXAMPLE

a) **The table and histogram show information about the masses of sharks in an aquarium. Use the histogram to find the missing entry in the table.**

Mass (m, kg)	Frequency
$750 < m \leq 850$	10
$850 < m \leq 900$	12
$900 < m \leq 1050$	51
$1050 < m \leq 1130$	

First, you'll need to add an extra column to the table to calculate frequency density:

Mass (m, kg)	Frequency	Frequency Density
$750 < m \leq 850$	10	$10 \div 100 = 0.1$
$850 < m \leq 900$	12	$12 \div 50 = 0.24$
$900 < m \leq 1050$	51	$51 \div 150 = 0.34$
$1050 < m \leq 1130$		

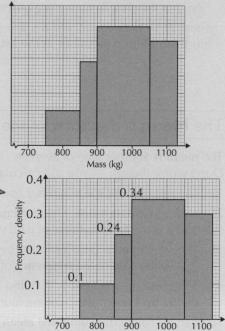

Now, use those frequency densities to work out the scale of the vertical axis of the graph:

This means you can read off the remaining bar — it has a frequency density of 0.3, and the table shows a class width of 80, so the missing frequency is $0.3 \times 80 = 24$.

$1050 < m \leq 1130$	**24**	**0.3**

b) **Estimate the number of sharks that weigh between 800 kg and 950 kg.**

Use the fact that frequency = frequency density × class width to estimate the frequency between 800 kg and 950 kg. The class widths are the bits of the classes you're interested in:

Between 800 and 850 kg, there are an estimated $0.1 \times (850 - 800) = 0.1 \times 50 = 5$ sharks.

Between 850 and 900 kg, there are 12 (you can read this from the table).

Between 900 and 950 kg, there are an estimated $0.34 \times (950 - 900) = 0.34 \times 50 = 17$ sharks.

This makes the estimate $5 + 12 + 17 = $ **34 sharks** that weigh between 800 kg and 950 kg.

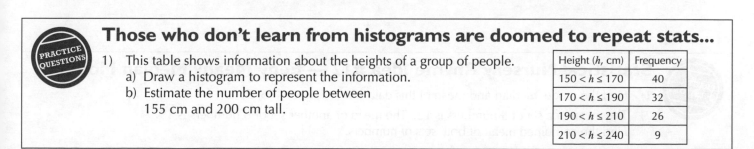

Those who don't learn from histograms are doomed to repeat stats...

PRACTICE QUESTIONS

1) This table shows information about the heights of a group of people.
   a) Draw a histogram to represent the information.
   b) Estimate the number of people between 155 cm and 200 cm tall.

Height (h, cm)	Frequency
$150 < h \leq 170$	40
$170 < h \leq 190$	32
$190 < h \leq 210$	26
$210 < h \leq 240$	9

Averages

The **Mode** and **Median** are **Averages**

You'll need to know all about the **different types** of average.

At A-Level, you may see these averages called 'measures of location'.

> The mode is the most commonly occurring data value.

If a set of data has two modes then it is called bimodal.

> The median is the middle value of the data, when the data is sorted in order of size.

Before you calculate the median, you need to put your **data** in **ascending order**.
Then you need to work out the **middle position**.

1) If you have an **odd** number of data values, then the **middle** point will be a **data value**.
2) If you have an **even** number of data values, the **middle** will be **between** two data values.
   So your **median** will be **halfway between** those values.

EXAMPLE

Find the median of this set of data: 8, 12, 4, 6, 13, 7, 8, 4.

This data is bimodal — the modes are 4 and 8.

First, put the data in ascending order: 4, 4, 6, 7, 8, 8, 12, 13.
There are 8 data values and $8 \div 2 = 4$. So the median will be halfway between the
fourth and fifth data values in the list. These values are 7 and 8, so the median is **7.5**.

Jane wanted to be a co-median, but her jokes were too average.

The **Mean** is the **Sum** of the **Data Divided** by the **Total Frequency**

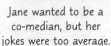

The **mean** is another type of **average**. At A-Level you'll use some **new notation** for it.
Don't worry though — you still calculate it like you did at GCSE.

1) The notation for the mean is $\bar{x}$ (said 'x-bar'). The **formula** is given as: $\bar{x} = \dfrac{\sum x}{n}$ or $\dfrac{\sum fx}{\sum f}$
2) Each x is a **data value**, f is the **frequency** of each x
   and n is the **total number** of data values.
3) The $\sum$ (called sigma) just means you **add things together**. So $\sum x$ is just the **sum** of all the **data values** x.
   $\sum f$ is the **sum** of the **frequency** of the data values — it's the same as saying '**the total number**' of data values.
4) So to find the mean of 1, 7, 3 and 5, **add them all together** to get **16** and then
   divide by 4 (because there are four values). This gives a mean of **4**.

You can calculate the **combined mean** of **two data** sets using this **formula**: $\bar{x} = \dfrac{n_1 \bar{x_1} + n_2 \bar{x_2}}{n_1 + n_2}$
All you need to know is the **mean** of each data set $\bar{x_1}$ and $\bar{x_2}$,
and the **size** of each data set n_1 and n_2.

EXAMPLE

**The mean of a set of 5 numbers is 20. The mean of another set of 7 numbers is 13.
Find the combined mean of both sets of numbers.**

Plug these numbers into the combined mean formula:

$\bar{x} = \dfrac{5 \times 20 + 7 \times 13}{5 + 7} = \dfrac{191}{12} = $ **15.92 (2 d.p.)**

PRACTICE QUESTIONS

Statistical Nursery Rhyme Idea #12 — Eeny Meany Median Mode...

1) Find the mode, median and mean of this data set: 1, 7, 8, 5, 4, 6, 5.
2) The mean of a set of 5 numbers is 12. The mean of another set of 6 numbers is 15.
   Find the combined mean of both sets of numbers.

Averages

Estimate Averages for Grouped Data

1) The **modal class** is the class with the highest **frequency density**.

2) You can also find the **class containing the median** — at A-Level you'll learn how to estimate the median.

3) To **estimate the mean**, you'll need to add extra **columns** — one for the **mid-point** of each class and another for **frequency × mid-point**.

4) Because you don't know **exact** values when you work with **grouped data**, these averages are only **estimations**.

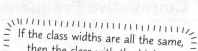

If the class widths are all the same, then the class with the highest frequency density will just be the class with the highest frequency.

EXAMPLES

This table shows information about the widths of cardboard boxes. State the modal class.

As all the class widths are the same, the modal class will be the class with the highest frequency: **40 < w ≤ 60 cm**.

Width (w, cm)	Frequency
20 < w ≤ 40	10
40 < w ≤ 60	11
60 < w ≤ 80	9

a) This table shows information about the length (to the nearest 10 mm) of 60 tree branches. Write down the class containing the median.

There are 60 data values, so the median is halfway between the 30th and 31st values. Count through the 'Frequency' column — both these values lie in the third class, so the class containing the median is **110-160 mm**.

Length (mm)	Frequency
0-50	5
60-110	7
110-160	22
160-210	16
210-260	10

b) Calculate an estimate for the mean length.

You'll need to add two extra columns to the table — one for each class's mid-point (x) and one for the value of the frequency multiplied by the class's mid-point (fx). You also need to add a total row.

Each class's mid-point is halfway between the upper and lower class boundaries. Be careful with the first one though — the upper class boundary is 55 and the lower class boundary is 0, so the mid-point is (0 + 55) ÷ 2 = 27.5.

Length (mm)	Frequency (f)	x	fx
0-50	5	27.5	137.5
60-110	7	85	595
110-160	22	135	2970
160-210	16	185	2960
210-260	10	235	2350
Total	60	–	9012.5

So the total frequency ($\sum f$) is 60 and the total of the frequencies multiplied by the mid-points ($\sum fx$) is 9012.5.

NEW CONTENT

Put these into the formula for the mean: $\bar{x} = 9012.5 \div 60 = \textbf{150.21}$ **(2 d.p.)**.

Learn this and become a lean, mean, average estimating machine...

PRACTICE QUESTIONS

1) This table shows information about the masses of pieces of fruit picked in an orchard (to the nearest 10 g).
   a) Find the class containing the median.
   b) Calculate an estimate for the mean weight.

Mass (g)	Frequency
0-40	6
50-90	11
100-140	8
150-190	5

Cumulative Frequency

| **Cumulative Frequency** is the **Running Total** of the **Frequency** |

1) You can draw a graph of the cumulative frequency of a data set, which lets you estimate the **median** and the **upper** and **lower quartiles** for the data.

2) The **upper quartile** is the value **75%** of the way through an **ordered set of data**.
   The **lower quartile** is the value **25%** of the way through an **ordered set of data**.

3) The **interquartile range** is the **difference** between the **upper** and **lower quartiles**.

EXAMPLE

a) The table below shows information about the masses of 100 puppies. Draw a cumulative frequency graph for the data.

The first thing you need to do is add a third column to the table for cumulative frequency.

To draw the graph, plot cumulative frequency up the vertical axis, and mass along the horizontal axis.

Start by plotting a cumulative frequency of zero at the lowest value in the first class — no puppies weigh less than 50 g.

Plot the rest of the points using the highest value of each class and the cumulative frequency.

Mass (m, g)	Frequency	Cumulative Frequency
$50 \leq m < 150$	5	**5**
$150 \leq m < 210$	7	$5 + 7 = $ **12**
$210 \leq m < 270$	16	$12 + 16 = $ **28**
$270 \leq m < 330$	25	$28 + 25 = $ **53**
$330 \leq m < 410$	31	$53 + 31 = $ **84**
$410 \leq m < 460$	11	$84 + 11 = $ **95**
$460 \leq m < 500$	5	$95 + 5 = $ **100**

b) Use your graph to estimate the median and interquartile range of the weights.

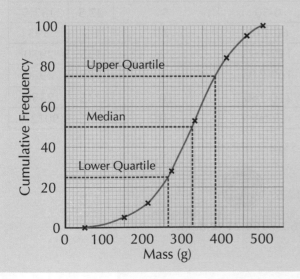

To find the median, go halfway up the side to 50 and read off the value from the curve: so an estimate of the **median** is **325 g**.

To find the interquartile range, you need to find the upper and lower quartiles first.
The upper quartile is 75% of the way through the data — so go across from 75 and read off the value from the curve: you get 380 g.
The lower quartile is 25% of the way through the data — so go across from 25 and read off the value from the curve: you get 260 g.
So an estimate of the **interquartile range** is 380 g – 260 g = **120 g**.

PRACTICE QUESTIONS

Interturtle range — about a flipper and a half...

1) The table on the right shows information about some lengths of copper wire.
   a) Draw a cumulative frequency graph for the data.
   b) Use your graph to estimate the median and the interquartile range of the lengths.

Length (l, mm)	Frequency
$300 \leq l < 350$	6
$350 \leq l < 410$	10
$410 \leq l < 470$	24
$470 \leq l < 520$	14
$520 \leq l < 600$	6

Probability

Remember these **Important Probability Facts**

You'll meet trickier probability in A-Level Maths, and build on your knowledge in new topics like **probability distributions**. So make sure you're **happy with the basics** here.

1) Finding the **probability** of something means working out **how likely** it is to happen. Your answer should always be between **0** (**impossible**) and **1** (**certain**).

2) For any probability experiment, the **different things** that can happen are called **outcomes**. E.g. if you flip a coin, the outcomes are 'heads' and 'tails'. The set of **all possible outcomes** is called the **sample space**.

3) The thing you want to find the probability of is often called an '**event**'. An event **matches one or more** of the possible **outcomes**, e.g. if you roll a dice, one event is 'roll a number less than 3' — which matches the outcomes '1' and '2'. You can write 'the probability of an event' as '**P(event)**' for short.

The probability of Hector being a good boy is 1.

4) There's a **handy formula** you can use to work out the probability of an event when **all** the possible outcomes are **equally likely**:

$$P(\text{event}) = \frac{\text{Number of outcomes where event happens}}{\text{Total number of possible outcomes}}$$

5) Another useful fact is that probability always **adds up to 1**. So for a set of events where **only one** of them can happen at a time, the probabilities of **all** the possible events **add up to 1**.

You can **Find** Probabilities from **Two-Way Tables**

You might be asked to work out probabilities using **information** in a **two-way frequency table**. The numbers in the table tell you how many **outcomes** match different **events**, and you can find the **total number** of outcomes in the **bottom right hand corner**.

EXAMPLE

a) **A bag contains 30 balls — coloured red, purple, or yellow, and labelled A or B. The numbers in each category are shown in the table. Find the probability that a ball selected at random is purple.**

	Red	Purple	Yellow	Total
A	7	3	4	14
B	3	5	8	16
Total	10	8	12	30

There are 8 purple balls in total, so there are 8 outcomes where the event 'ball is purple' happens. The total number of possible outcomes is the total number of balls, which is 30.

So using the probability formula above: P(ball is purple) = $\frac{8}{30} = \frac{4}{15}$

In the question, 'at random' tells you the outcomes are all equally likely, so you can use the formula.

b) **Find the probability that a ball selected at random is red or labelled A.**

You need the number of red balls, plus the number of purple and yellow balls labelled A — that's 10 + 3 + 4 = 17. So P(ball is red or labelled A) = $\frac{17}{30}$

Chances are you almost certainly didn't hate this page — right?

1) I roll a fair, six-sided dice. Find the probability that I roll: a) 1 or 2   b) higher than 2.

2) Using the two-way table in the example above, find the probability that a randomly selected ball is yellow and labelled A.

Probability

A **Venn Diagram** Shows How Things are **Grouped**

Venn diagrams show how a **collection of objects** is **split up** into different **groups**.

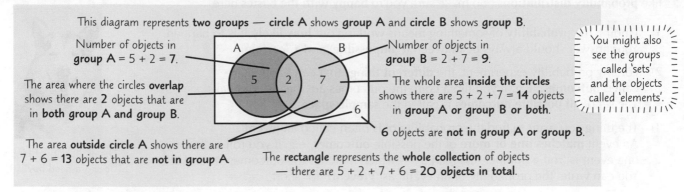

This diagram represents **two groups** — circle A shows **group A** and circle B shows **group B**.

Number of objects in **group A** = 5 + 2 = **7**.

The area where the circles **overlap** shows there are **2** objects that are in **both group A and group B**.

The area **outside circle A** shows there are 7 + 6 = **13** objects that are **not in group A**.

Number of objects in **group B** = 2 + 7 = **9**.

The whole area **inside the circles** shows there are 5 + 2 + 7 = **14** objects in **group A or group B or both**.

6 objects are **not in group A or group B**.

The **rectangle** represents the **whole collection** of objects — there are 5 + 2 + 7 + 6 = **20 objects in total**.

You might also see the groups called 'sets' and the objects called 'elements'.

Finding Probabilities from Venn Diagrams

You've used Venn diagrams to **find probabilities** at GCSE, and you'll use them again at A-Level. They're really useful for solving probability problems — the labels show the **number of outcomes** (or the **probabilities**) matching the **different events** that can happen.

EXAMPLE

This Venn diagram shows the number of children at a primary school who go to baking club (B) and singing club (S). Find the probability that a randomly selected child from the school goes to baking club.

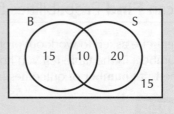

The event 'child goes to baking club' is represented by circle B. So there are 15 + 10 = 25 outcomes matching this event. And there are 15 + 10 + 20 + 15 = 60 children in total.

So using the formula from p.57 you get: P(child goes to baking club) = $\frac{25}{60}$ = $\frac{5}{12}$

Here's a trickier example with **three** circles — and this time the diagram is **labelled** with the **proportion** in each area, rather than the number.

EXAMPLE

A group of students were asked which of biology (B), chemistry (C) and physics (P) they like. The diagram shows the proportion who like each subject. If one student is selected at random, find the probability that they like physics, but not chemistry.

'Likes physics, but not chemistry' is shown by the area that's in P, but not in C.

The probability of the student being in this area is equal to the proportion in the area.

So add up the numbers to get: P(likes physics, but not chemistry) = 0.05 + 0.05 = **0.1**

You can have a cup of tea venn you've done these questions...

1) Out of 40 customers at a book shop, 22 bought fiction books, 7 bought travel books, and 2 bought both fiction and travel books. Draw a Venn diagram to represent the 40 customers.

2) Using the Venn diagram in the second example above, find the probability that a randomly selected student likes biology or chemistry or both.

Laws of Probability

The **Addition Law** Tells You **P(A or B)**

This law should be familiar from GCSE. You'll see how to write it using set notation in Year 2 of A-Level Maths.

Say you have **two events**, A and B. You can find the probability of **event A or event B or both** happening using the **addition law.**

$$P(A \text{ or } B) = P(A) + P(B) - P(A \text{ and } B)$$

If the events A and B **can't happen** at the same time, they're called **mutually exclusive** events. For mutually exclusive events A and B, **P(A and B) = 0**, so the addition law becomes:

$$P(A \text{ or } B) = P(A) + P(B)$$

You can **show** whether or not two events are **mutually exclusive**. All you have to do is to show whether or not P(A and B) **equals zero**.

EXAMPLE

P(A) = 0.16, P(B) = 0.5 and P(A or B) = 0.62.
Show whether or not events A and B are mutually exclusive.

Rearrange the addition law to find P(A and B):   P(A or B) = P(A) + P(B) – P(A and B)
⇒ P(A and B) = P(A) + P(B) – P(A or B) = 0.16 + 0.5 – 0.62 = 0.04

P(A and B) ≠ 0, so events A and B are **not mutually exclusive**.

The **Product Law** Tells You **P(A and B)**

You need a different form of the law for dependent events — it's covered in Year 2 of A-Level.

If one of events A and B happening **doesn't affect** the **probability** of the other happening, **events A and B** are **independent**. This means you can use the **product law** for **independent** events: $P(A \text{ and } B) = P(A) \times P(B)$

EXAMPLE

Sid buys one piece of fruit each day. The probability he buys an apple, P(A), is always 0.4, independent of any other day. Find the probability that he buys an apple on exactly one of the next two days.

This can happen in one of two ways: either Sid buys an apple on the first day and not on the second day, or he doesn't buy an apple on the first day, but does on the second day.

P(A) = 0.4, so P(not A) = 1 – 0.4 = 0.6, since probabilities add up to 1.

The days are independent, so you can multiply the probabilities using the product law to find P(A and 'not A') and P('not A' and A):
P(A and 'not A') = 0.4 × 0.6 = 0.24 and P('not A' and A) = 0.6 × 0.4 = 0.24.

These two options can't both happen — either one happens or the other happens, so you add the probabilities using the addition rule for mutually exclusive events:
P(one apple in two days) = P('A and not A' or 'not A and A')
= P(A and 'not A') + P('not A' and A) = 0.24 + 0.24 = **0.48**.

Like with mutually exclusive events, you can **show** whether or not two events are **independent**. You just need to show whether or not **P(A) × P(B) = P(A and B)** — see Question 2 below.

Probability law #3 — learn probability laws 1 and 2, or else...

1) On any given day, the probabilities that Martha and Sahil play tennis are 0.3 and 0.4 respectively. Find the probability that, on a given day, Martha or Sahil plays tennis.
2) If P(A) = 0.6, P(B) = 0.5 and P(A and B) = 0.3, show that events A and B are independent.

Tree Diagrams

Tree Diagrams Show Probabilities for a Sequence of Events

To **draw** a tree diagram representing two 'trials':

1) Draw a set of branches for the first trial, with a branch for each result.
2) At the end of each branch, draw a set of branches for the second trial.
3) Label each branch with the probability of that result — probabilities on each set of branches should add up to 1.
4) To find the probability of a sequence of events, multiply along the branches. The probabilities of all the sequences added together should always be 1.

A 'trial' is just something like rolling a dice, or selecting an object at random. For three trials, just repeat step 2.

EXAMPLE

A fair, six-sided dice is rolled three times. Use a tree diagram to find the probability that exactly two even numbers are rolled.

There are three trials and you can define two results for each trial — 'even number' (E), and 'odd number' (O), where each result has a probability of $\frac{1}{2}$.

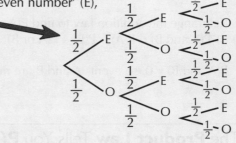

There are three sequences that give two even numbers — (E, E, O), (E, O, E) and (O, E, E). Multiplying along the branches, the probability of each sequence $= \frac{1}{2} \times \frac{1}{2} \times \frac{1}{2} = \frac{1}{8}$.

You want to find the probability of (E, E, O) or (E, O, E) or (O, E, E), so add the probabilities:

P(exactly two even numbers) = P(E, E, O) + P(E, O, E) + P(O, E, E) = $\frac{1}{8} + \frac{1}{8} + \frac{1}{8} = \frac{3}{8}$

Tree Diagrams can show Dependent Events

Events are **dependent** if one event happening **affects the probability** of the other happening. When you're drawing a tree diagram for dependent events, you need to **change** the **probabilities** on the **second** set of branches, **depending** on the result in the **first** set.

EXAMPLE

A cake tin contains 6 chocolate muffins and 4 blueberry muffins. Two muffins are taken at random without replacement. Using a tree diagram, find the probability that at least one chocolate muffin is taken.

There are two trials, each with results 'chocolate' (C) and 'blueberry' (B). The probabilities for the second trial change because the first muffin isn't replaced. There are now 9 muffins left in total, and the number of each flavour depends on the first selection.

P(at least 1 choc) + P(no choc) = 1.

P(at least one chocolate) = 1 − P(no chocolate)

= 1 − P(two blueberry) = $1 - \left(\frac{4}{10} \times \frac{3}{9}\right) = 1 - \frac{2}{15} = \frac{13}{15}$

Tree diagram branches:
- $\frac{6}{10}$ C → $\frac{5}{9}$ C, $\frac{4}{9}$ B
- $\frac{4}{10}$ B → $\frac{6}{9}$ C, $\frac{3}{9}$ B

Sustainable stats — for every page I write, I draw a tree diagram...

1) Claire eats either pizza or curry every Friday. The probability that she has pizza is 0.25 if she had pizza the previous Friday, and 0.6 if she had curry the previous Friday. Given that Claire has curry this Friday, find the probability that she has pizza on both of the next two Fridays.

Answers

Page 1 — Diagnostic Test

1   4, –10, 205 and 0 are integers.

2     $5.\dot{9}$, $\frac{1}{5}$, –6, $\sqrt{4}$, 13.978 and 2.1 are rational.

      π and $\sqrt{7}$ are irrational.

3   a) $\frac{2}{15}$

    b) $\frac{1}{4}$

    c) $\frac{11}{12}$

    d) $\frac{51}{35}$

4   a) x^9
    b) $2y^2$
    c) 1
    d) $32n^{10}$

5     $\frac{1}{25}$

6   a) $\frac{9}{16}$
    b) 4
    c) 4
    d) $\frac{1}{6}$

7   a) $x^2 - 2x - 24$
    b) $x^2 + 10x + 25$
    c) $2x^2 + 5x - 3$
    d) $x^3 + 2x^2 - 19x - 20$

8   a) $5(x + 4)$
    b) $3a(1 + 4b)$
    c) $(x + 2)(x - 2)$
    d) $9(x + 2)(x - 2)$
    e) $(x + \sqrt{5})(x - \sqrt{5})$

9   a) $\sqrt{6}$
    b) 5
    c) $\sqrt{5}$
    d) $4\sqrt{3}$
    e) $8 + 2\sqrt{7}$

10 a) $\frac{3\sqrt{2}}{2}$

    b) $\frac{\sqrt{10}}{4}$

    c) $\frac{6 - 2\sqrt{6}}{3}$

    d) $\frac{\sqrt{2} + \sqrt{10}}{-4}$

11 a) $x = 2$
    b) $x = 10$
    c) $x = -5$
    d) $x = -3$ or $x = 3$

12 a) $x = \frac{y - c}{m}$

    b) $x = \frac{5y - 2}{3}$

    c) $x = \pm\sqrt{\dfrac{y - 1}{2z}}$

    d) $x = \frac{2y + 1}{y - 3}$

13 a) $x = 2$ or $x = 1$
    b) $x = -5$ or $x = -1$
    c) $x = 2.5$ or $x = -1$
    d) $x = \frac{4}{3}$ or $x = 3$

14 a) $x = 2.32$ or $x = -4.32$
    b) $x = 2.69$ or $x = -0.19$

15 a) $x = 2 + \sqrt{6}$ or $x = 2 - \sqrt{6}$
    b) $x = -1 + \dfrac{3}{\sqrt{2}}$ or $x = -1 - \dfrac{3}{\sqrt{2}}$

16 a) $x^2 + 6x + 8 = (x + 3)^2 - 1$
    b)

17 a) $3ab^2$
    b) $\frac{1}{8y}$
    c) $\frac{x - 4}{x - 5}$

18 a) $6ab$
    b) $\frac{x - 1}{3}$
    c) 27
    d) $\frac{5x^2 - x - 3}{x^2(x + 1)}$

19 a) $x \le -1$
    b) $x < 8$
    c) $-3 < x < 3$
    d) $x \le -2$ or $x \ge 2$
    e) $x \le 1$ or $x \ge 5$

20

21 a) $x = 2$, $y = -2$
    b) $x = 11$, $y = 16$
    c) $x = -3$, $y = 12$ or $x = 5$, $y = 28$
    d) $x = 0$, $y = -2$ or $x = 3$, $y = 4$

22   Take three consecutive odd numbers:
    $2n + 1$, $2n + 3$ and $2n + 5$, where n is an integer.
    $2n + 1 + 2n + 3 + 2n + 5 = 6n + 9 = 3(2n + 3)$
    The sum of three consecutive odd numbers can be written as $3x$, where $x = 2n + 3$.
    Therefore it is a multiple of 3.

23   E.g. Let $x = 3$ and $y = -1$. So $xy = -3 \Rightarrow xy < y$.
    So Naveen is wrong.

Answers

24 a) 3

b) $fg(x) = \dfrac{x+2}{3}$

c) $f^{-1}(x) = 3x - 5$

25 Gradient = –0.5, y-intercept = 2

26 a) $y = 2x - 8$

b) $3\sqrt{5}$

27 a) $y = 2x - 4$

b) $y = -\dfrac{1}{2}x + 2$

28

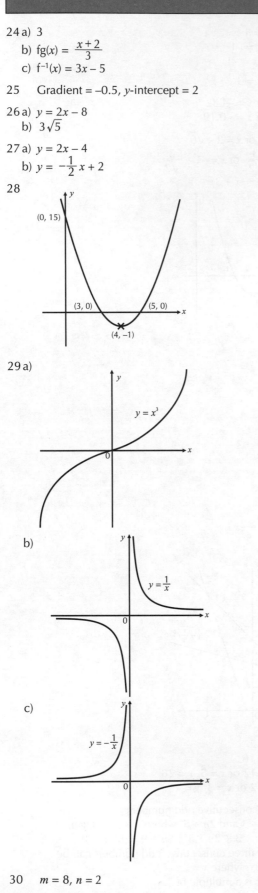

29 a)

b)

c)

30 $m = 8$, $n = 2$

31 $3x + 4y - 25 = 0$

32 a)

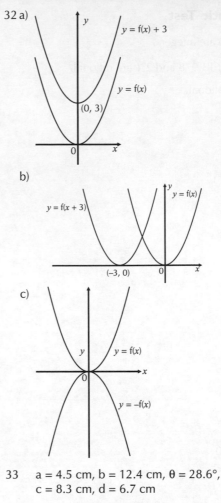

b)

c)

33 $a = 4.5$ cm, $b = 12.4$ cm, $\theta = 28.6°$, $c = 8.3$ cm, $d = 6.7$ cm

34 $a = 3.8$ cm, $b = 40.3°$, $c = 7.2$ cm, $d = 62.7°$

35 a) $\mathbf{a} + \mathbf{b}$

b) $\dfrac{1}{2}\mathbf{a}$

c) $-\mathbf{a} - \dfrac{1}{2}\mathbf{b}$

d) $\dfrac{1}{2}\mathbf{a} - \dfrac{1}{2}\mathbf{b}$

e) $\mathbf{a} - \dfrac{1}{2}\mathbf{b}$

f) $-\mathbf{a} - \dfrac{1}{2}\mathbf{b}$

36 $\overrightarrow{AB} = 2(3\mathbf{a} - \mathbf{b}) - 2(2(\mathbf{a} - \mathbf{b}))$
     $= 6\mathbf{a} - 2\mathbf{b} - 4\mathbf{a} + 4\mathbf{b} = 2\mathbf{a} + 2\mathbf{b} = 2(\mathbf{a} + \mathbf{b})$

   $\overrightarrow{MN} = 3\mathbf{a} - \mathbf{b} - 2(\mathbf{a} - \mathbf{b}) = 3\mathbf{a} - \mathbf{b} - 2\mathbf{a} + 2\mathbf{b} = \mathbf{a} + \mathbf{b}$

   $\overrightarrow{AB} = 2\overrightarrow{MN} \Rightarrow \overrightarrow{AB}$ and $\overrightarrow{MN}$ are parallel.

37 First assign a unique number between 1 and 200 to every member of the population. Then create a list of 20 random numbers between 1 and 200. Finally, match the random numbers to members of the population.

38 a) See histogram in part c).

b)

Weight (w, in grams)	Frequency
$0 < w \leq 100$	50
$100 < w \leq 150$	100
$150 < w \leq 200$	150
$200 < w \leq 250$	**90**

c)

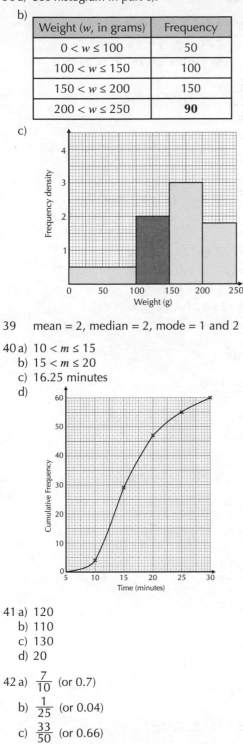

39   mean = 2, median = 2, mode = 1 and 2

40 a) $10 < m \leq 15$
   b) $15 < m \leq 20$
   c) 16.25 minutes
   d)

41 a) 120
   b) 110
   c) 130
   d) 20

42 a) $\frac{7}{10}$ (or 0.7)

   b) $\frac{1}{25}$ (or 0.04)

   c) $\frac{33}{50}$ (or 0.66)

   d) $\frac{2}{25}$ (or 0.08)

   e) $\frac{3}{10}$ (or 0.3)

43 a) $\frac{5}{54}$
   b)

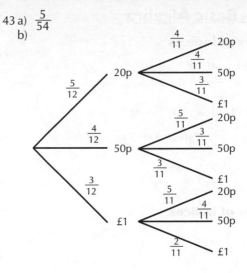

Probability of a 50p coin and a £1 coin $= \frac{2}{11}$

Section 1 — Number

Page 6 — Types of Number

1 a) **Rational** — it's a terminating decimal.
   b) **Rational** — it's a recurring decimal.
   c) **Rational** — $\sqrt{64}$ = 8 which is an integer.
   d) **Irrational** — 3 is not a square number.
   e) **Irrational** — π is irrational

2 a) **True** — any integer can be written as a fraction over 1.
   b) **False** — recurring decimals are rational.

Page 7 — Fractions

1 a) $2\frac{2}{3} \times \frac{1}{4} = \frac{8}{3} \times \frac{1}{4} = \frac{2}{3} \times \frac{1}{1} = \mathbf{\frac{2}{3}}$

   b) $5\frac{1}{3} \div 2\frac{1}{4} = \frac{16}{3} \div \frac{9}{4} = \frac{16}{3} \times \frac{4}{9} = \mathbf{\frac{64}{27}}$

   c) $\frac{3}{4} + \frac{1}{3} = \frac{9}{12} + \frac{4}{12} = \mathbf{\frac{13}{12}}$

   d) $1\frac{1}{6} + 2\frac{1}{2} = \frac{7}{6} + \frac{5}{2} = \frac{7}{6} + \frac{15}{6} = \frac{22}{6} = \mathbf{\frac{11}{3}}$

   e) $1\frac{3}{7} - \frac{2}{9} = \frac{10}{7} - \frac{2}{9} = \frac{90}{63} - \frac{14}{63} = \mathbf{\frac{76}{63}}$

   f) $\frac{1}{2} + \frac{2}{3} + \frac{3}{4} = \frac{6}{12} + \frac{8}{12} + \frac{9}{12} = \mathbf{\frac{23}{12}}$

   g) $\left(\frac{7}{6} - \frac{1}{7}\right) \div \frac{1}{2} = \left(\frac{49}{42} - \frac{6}{42}\right) \div \frac{1}{2} = \frac{43}{42} \div \frac{1}{2} = \frac{43}{42} \times \frac{2}{1} = \mathbf{\frac{43}{21}}$

   h) $\frac{3}{2} - \frac{1}{9} + 2\frac{1}{3} = \frac{3}{2} - \frac{1}{9} + \frac{7}{3} = \frac{3}{2} - \frac{1}{9} + \frac{21}{9}$
   $= \frac{3}{2} + \frac{20}{9} = \frac{27}{18} + \frac{40}{18} = \mathbf{\frac{67}{18}}$

Answers

Section 2 — Basic Algebra

Page 8 — Laws of Indices

1 a) $b^5 \times b^6 = b^{(5+6)} = \boldsymbol{b^{11}}$

b) $a^9 \times a \times b^5 = a^{(9+1)} \times b^5 = \boldsymbol{a^{10}b^5}$

c) $c^5 \div c^2 = c^{(5-2)} = \boldsymbol{c^3}$

d) $9y^{10} \div 3y^{-2} = (9 \div 3) \times y^{(10-(-2))} = \boldsymbol{3y^{12}}$

e) $x^2 \times x^3 \div x^4 = x^{(2+3-4)} = \boldsymbol{x}$

f) $z^3 \times (y+2)^5 \times z \div (y+2)^2 = z^{(3+1)} \times (y+2)^{(5-2)}$
$= \boldsymbol{z^4(y+2)^3}$

g) $a^{(x+2)} \times a^{2x} = a^{(x+2)+2x} = \boldsymbol{a^{3x+2}}$

h) $x^{-2}y^5 \times x^5y^2 = x^{(-2+5)}y^{(5+2)} = \boldsymbol{x^3y^7}$

Page 9 — Laws of Indices

1 a) $a^0 \div b^{-2} = 1 \div \dfrac{1}{b^2} = \boldsymbol{b^2}$

b) $(4^x)^x = \boldsymbol{4^{x^2}}$

c) $3^m \times 9^2 = 3^m \times (3^2)^2 = 3^m \times 3^4 = \boldsymbol{3^{m+4}}$

d) $\left(\dfrac{1}{a}\right)^2 \times a^{-3} = a^{-2} \times a^{-3} = a^{-2-3} = \boldsymbol{a^{-5}}$

e) $\left(\dfrac{1}{z^9}\right)^{\frac{1}{3}} = \dfrac{1^{\frac{1}{3}}}{(z^9)^{\frac{1}{3}}} = \boldsymbol{\dfrac{1}{z^3}}$

2 a) $4^{\frac{1}{2}} = \sqrt{4} = \boldsymbol{2}$

b) $27^{\frac{2}{3}} = (\sqrt[3]{27})^2 = (3)^2 = \boldsymbol{9}$

c) $2 \times 32^{\frac{3}{5}} = 2 \times (2^5)^{\frac{3}{5}} = 2 \times 2^3 = 2^4 = \boldsymbol{16}$

d) $\left(\dfrac{125}{8}\right)^{\frac{1}{3}} = \sqrt[3]{\dfrac{125}{8}} = \dfrac{\sqrt[3]{125}}{\sqrt[3]{8}} = \boldsymbol{\dfrac{5}{2}}$

e) $\left(\dfrac{25}{4}\right)^{-\frac{3}{2}} = \left(\dfrac{4}{25}\right)^{\frac{3}{2}} = \left(\sqrt{\dfrac{4}{25}}\right)^3 = \left(\dfrac{\sqrt{4}}{\sqrt{25}}\right)^3 = \left(\dfrac{2}{5}\right)^3 = \boldsymbol{\dfrac{8}{125}}$

f) $\left(\dfrac{16}{9}\right)^{-\frac{3}{2}} = \left(\dfrac{9}{16}\right)^{\frac{3}{2}} = \left(\sqrt{\dfrac{9}{16}}\right)^3 = \left(\dfrac{\sqrt{9}}{\sqrt{16}}\right)^3 = \left(\dfrac{3}{4}\right)^3 = \boldsymbol{\dfrac{27}{64}}$

3 $(5^{\frac{1}{4}})^2 \times (5^{\frac{2}{3}})^{-\frac{3}{4}} \div (5^{-1})^{-2} = 5^{\frac{2}{4}} \times 5^{(\frac{2}{3} \times -\frac{3}{4})} \div 5^{-1 \times -2}$
$= 5^{\frac{1}{2}} \times 5^{-\frac{1}{2}} \div 5^2$
$= 5^0 \div 5^2 = \boldsymbol{5^{-2}}$

Page 10 — Multiplying Out Brackets

1 a) $(y+3)(y-6) = y^2 - 6y + 3y - 18 = \boldsymbol{y^2 - 3y - 18}$

b) $(a-3)(b+4) = \boldsymbol{ab + 4a - 3b - 12}$

c) $(p-1)(p-2q)^2 = (p-1)(p-2q)(p-2q)$
$= (p-1)(p^2 - 4pq + 4q^2)$
$= p(p^2 - 4pq + 4q^2) + (-1)(p^2 - 4pq + 4q^2)$
$= \boldsymbol{p^3 - 4p^2q + 4pq^2 - p^2 + 4pq - 4q^2}$

d) $(s^2 + s + 2)(2s^2 - 2s + 4)$
$= s^2(2s^2 - 2s + 4) + s(2s^2 - 2s + 4) + 2(2s^2 - 2s + 4)$
$= (2s^4 - 2s^3 + 4s^2) + (2s^3 - 2s^2 + 4s) + (4s^2 - 4s + 8)$
$= \boldsymbol{2s^4 + 6s^2 + 8}$

Page 11 — Factorising

1 a) $20x^2 - 4x = \boldsymbol{4x(5x-1)}$

b) $8x^2y + 28xy^2 = \boldsymbol{4xy(2x+7y)}$

c) $3\pi a^2 + 4\pi ab + 2\pi a = \boldsymbol{\pi a(3a+4b+2)}$

d) $5x^2(x-1) - 2x(x-1) = \boldsymbol{x(x-1)(5x-2)}$

e) $x^2 - 9 = \boldsymbol{(x+3)(x-3)}$

f) $9x^2 - 25 = \boldsymbol{(3x+5)(3x-5)}$

g) $p^2 - 49q^2 = p^2 - (7q)^2 = \boldsymbol{(p+7q)(p-7q)}$

h) $v^2 - 7u^2 = \boldsymbol{(v + \sqrt{7}\,u)(v - \sqrt{7}\,u)}$

Page 12 — Surds

1 a) $2\sqrt{24} + 3\sqrt{96} = 2\sqrt{4 \times 6} + 3\sqrt{16 \times 6}$
$= 2\sqrt{4}\sqrt{6} + 3\sqrt{16}\sqrt{6}$
$= 4\sqrt{6} + 12\sqrt{6} = \boldsymbol{16\sqrt{6}}$

b) $\dfrac{\sqrt{120}}{\sqrt{15}\sqrt{2}} = \dfrac{\sqrt{120}}{\sqrt{30}} = \sqrt{\dfrac{120}{30}} = \sqrt{4} = \boldsymbol{2}$

c) $(1 + \sqrt{x})^2 - 2\sqrt{x} = (1 + \sqrt{x})(1 + \sqrt{x}) - 2\sqrt{x}$
$= 1 + 2\sqrt{x} + (\sqrt{x})^2 - 2\sqrt{x} = \boldsymbol{1 + x}$

2 $\dfrac{\sqrt{3}}{\sqrt{20}} = \dfrac{\sqrt{3}}{\sqrt{4}\sqrt{5}} = \dfrac{\sqrt{3}}{2\sqrt{5}} = \dfrac{\sqrt{3}\sqrt{5}}{2\sqrt{5}\sqrt{5}} = \dfrac{\sqrt{15}}{2(\sqrt{5})^2}$
$= \dfrac{\sqrt{15}}{2 \times 5} = \boldsymbol{\dfrac{\sqrt{15}}{10}}$

3 $(\sqrt{7} + \sqrt{12})^2 = (\sqrt{7} + \sqrt{12})(\sqrt{7} + \sqrt{12})$
$= (\sqrt{7})^2 + 2\sqrt{7}\sqrt{12} + (\sqrt{12})^2$
$= 7 + 2\sqrt{84} + 12$
$= 19 + 2\sqrt{4}\sqrt{21}$
$= \boldsymbol{19 + 4\sqrt{21}}$

Page 13 — Surds

1 a) $\dfrac{1}{1 - \sqrt{5}} = \dfrac{1 \times (1 + \sqrt{5})}{(1 - \sqrt{5})(1 + \sqrt{5})}$
$= \dfrac{1 + \sqrt{5}}{1 - (\sqrt{5})^2} = \dfrac{1 + \sqrt{5}}{1 - 5} = \boldsymbol{-\dfrac{1 + \sqrt{5}}{4}}$

b) $\dfrac{\sqrt{10}}{4 + \sqrt{40}} = \dfrac{\sqrt{10}}{4 + \sqrt{4}\sqrt{10}} = \dfrac{\sqrt{10}}{4 + 2\sqrt{10}}$
$= \dfrac{\sqrt{10}(4 - 2\sqrt{10})}{(4 + 2\sqrt{10})(4 - 2\sqrt{10})}$
$= \dfrac{4\sqrt{10} - 20}{4^2 - (2\sqrt{10})^2} = \dfrac{4\sqrt{10} - 20}{16 - 4 \times 10}$
$= \dfrac{4\sqrt{10} - 20}{-24} = \dfrac{\sqrt{10} - 5}{-6} = \boldsymbol{\dfrac{5 - \sqrt{10}}{6}}$

c) $\dfrac{1 + \sqrt{7}}{5 + \sqrt{7}} = \dfrac{(1 + \sqrt{7})(5 - \sqrt{7})}{(5 + \sqrt{7})(5 - \sqrt{7})}$
$= \dfrac{5 - \sqrt{7} + 5\sqrt{7} - (\sqrt{7})^2}{5^2 - (\sqrt{7})^2}$
$= \dfrac{5 + 4\sqrt{7} - 7}{25 - 7} = \dfrac{-2 + 4\sqrt{7}}{18} = \boldsymbol{\dfrac{2\sqrt{7} - 1}{9}}$

d) $\dfrac{2 + 2\sqrt{2}}{2 - 2\sqrt{2}} = \dfrac{(2 + 2\sqrt{2})(2 + 2\sqrt{2})}{(2 - 2\sqrt{2})(2 + 2\sqrt{2})}$
$= \dfrac{4 + 4\sqrt{2} + 4\sqrt{2} + (2\sqrt{2})^2}{4 - (2\sqrt{2})^2}$
$= \dfrac{4 + 8\sqrt{2} + 8}{4 - 8} = \dfrac{12 + 8\sqrt{2}}{-4} = \boldsymbol{-3 - 2\sqrt{2}}$

2 $\dfrac{4}{1 - 2\sin 60°} = \dfrac{4}{1 - 2\left(\frac{\sqrt{3}}{2}\right)}$
$= \dfrac{4}{1 - \sqrt{3}} = \dfrac{4(1 + \sqrt{3})}{(1 - \sqrt{3})(1 + \sqrt{3})}$
$= \dfrac{4 + 4\sqrt{3}}{1 - (\sqrt{3})^2} = \dfrac{4 + 4\sqrt{3}}{1 - 3}$
$= \dfrac{4 + 4\sqrt{3}}{-2} = \boldsymbol{-2 - 2\sqrt{3}}$

Answers

Page 14 — Solving Equations

1 a) $4(2x - 3) = 7x \Rightarrow 8x - 12 = 7x \Rightarrow 8x = 7x + 12 \Rightarrow x = \mathbf{12}$

b) $3(x + 14) = x + 12 \Rightarrow 3x + 42 = x + 12$
$$\Rightarrow 2x = -30 \Rightarrow x = \mathbf{-15}$$

c) $\dfrac{b - 7}{3} + \dfrac{b + 1}{5} = -1 \Rightarrow 5(b - 7) + 3(b + 1) = -15$
$$\Rightarrow 5b - 35 + 3b + 3 = -15$$
$$\Rightarrow 8b - 32 = -15 \Rightarrow 8b = 17$$
$$\Rightarrow b = \mathbf{\dfrac{17}{8}}$$

d) $\dfrac{q(q + 7)}{7} - q = 4 - q^2 \Rightarrow q(q + 7) - 7q = 28 - 7q^2$
$$\Rightarrow q^2 + 7q - 7q = 28 - 7q^2$$
$$\Rightarrow 8q^2 = 28 \Rightarrow q^2 = \dfrac{28}{8} = \dfrac{7}{2}$$
$$\Rightarrow q = \mathbf{\pm\sqrt{\dfrac{7}{2}}}$$

2 The area of the whole square will be x^2.
If the area of $\frac{1}{4}$ of the square is 25, then $\dfrac{x^2}{4} = 25$
$\Rightarrow x^2 = 100 \Rightarrow x = \pm 10$.
So the sides are **10 cm**.

Page 15 — Rearranging Formulas

1 a) $b(a + 2) = 4 \Rightarrow ab + 2b = 4 \Rightarrow ab = 4 - 2b \Rightarrow a = \mathbf{\dfrac{4}{b} - 2}$

b) $f = \dfrac{9}{5}c + 32 \Rightarrow 5f = 9c + 160 \Rightarrow 9c = 5f - 160$
$$\Rightarrow c = \mathbf{\dfrac{5f - 160}{9}}$$

2 a) $y = \dfrac{2x^2 - 3}{x^2 - 1} \Rightarrow y(x^2 - 1) = 2x^2 - 3$
$$\Rightarrow x^2 y - y = 2x^2 - 3$$
$$\Rightarrow x^2 y - 2x^2 = y - 3$$
$$\Rightarrow (y - 2)x^2 = y - 3$$
$$\Rightarrow x^2 = \dfrac{y - 3}{y - 2} \Rightarrow x = \mathbf{\pm\sqrt{\dfrac{y - 3}{y - 2}}}$$

b) $s = \dfrac{\sqrt{t + u}}{u} \Rightarrow s^2 = \dfrac{t + u}{u^2} \Rightarrow s^2 u^2 = t + u \Rightarrow t = \mathbf{s^2 u^2 - u}$

Section 3 — Quadratic Equations

Page 16 — Factorising Quadratics

1 a) $x^2 + 3x + 2 = 0 \Rightarrow (x + 1)(x + 2) = 0 \Rightarrow x = \mathbf{-1}$ and $x = \mathbf{-2}$

b) $x^2 + 8x + 7 = 0 \Rightarrow (x + 1)(x + 7) = 0 \Rightarrow x = \mathbf{-1}$ and $x = \mathbf{-7}$

c) $x^2 = 2(7x - 20) \Rightarrow x^2 - 14x + 40 = 0 \Rightarrow (x - 4)(x - 10) = 0$
$$\Rightarrow x = \mathbf{4}$$ and $x = \mathbf{10}$

d) $x^2 - x = 6 \Rightarrow x^2 - x - 6 = 0 \Rightarrow (x + 2)(x - 3) = 0$
$$\Rightarrow x = \mathbf{-2}$$ and $x = \mathbf{3}$

e) $7x = x^2 + 10 \Rightarrow x^2 - 7x + 10 = 0 \Rightarrow (x - 2)(x - 5) = 0$
$$\Rightarrow x = \mathbf{2}$$ and $x = \mathbf{5}$

f) Multiply everything by 2 to get $x^2 + 4x - 12 = 0$.
$x^2 + 4x - 12 = 0 \Rightarrow (x - 2)(x + 6) = 0 \Rightarrow x = \mathbf{2}$ and $x = \mathbf{-6}$

g) Multiply everything by x to get $x^2 - 4x - 12 = 0$.
$x^2 - 4x - 12 = 0 \Rightarrow (x + 2)(x - 6) = 0 \Rightarrow x = \mathbf{-2}$ and $x = \mathbf{6s}$

h) $2x^2 - 2x - 4 = 0 \Rightarrow x^2 - x - 2 = 0$
$\Rightarrow (x - 2)(x + 1) = 0 \Rightarrow x = \mathbf{2}$ and $x = \mathbf{-1}$

2 $x^2 - 2xz + z^2 = (x - z)(x - z) = \mathbf{(x - z)^2}$

3 $-m^2 + 13m - 30 = 0 \Rightarrow m^2 - 13m + 30 = 0$
$\Rightarrow (m - 3)(m - 10) \Rightarrow m = 3$ and 10. So the temperature
was 0°C after **3 minutes** and after **10 minutes.**

Page 17 — Factorising Quadratics

1 a) $2x^2 + 9x + 9 = 0 \Rightarrow (2x + 3)(x + 3) = 0$
$$\Rightarrow x = \mathbf{-\dfrac{3}{2}}$$ and $x = \mathbf{-3}$

b) $5x^2 + 13x + 6 = 0 \Rightarrow (5x + 3)(x + 2) = 0$
$$\Rightarrow x = \mathbf{-\dfrac{3}{5}}$$ and $x = \mathbf{-2}$

c) $2x^2 = x + 10 \Rightarrow 2x^2 - x - 10 = 0 \Rightarrow (2x - 5)(x + 2)$
$$\Rightarrow x = \mathbf{\dfrac{5}{2}}$$ and $x = \mathbf{-2}$

d) $3x + \dfrac{21}{x} = 16$, so multiply through by x:
$3x^2 + 21 = 16x \Rightarrow 3x^2 - 16x + 21 = 0$
$$\Rightarrow (3x - 7)(x - 3) = 0$$
$$\Rightarrow x = \mathbf{\dfrac{7}{3}}$$ and $x = \mathbf{3}$.

2 The plane hits the ground when $-4t^2 + 2t + 2 = 0$.
Divide through by -2: $2t^2 - t - 1 = 0$
$\Rightarrow (2t + 1)(t - 1) = 0 \Rightarrow t = 1$ or $t = -\dfrac{1}{2}$.
t can't be negative, so the plane hits the ground
1 second after being thrown.

Page 18 — The Quadratic Formula

1 a) $4x^2 - 7x + 1 = 0 \Rightarrow a = 4, b = -7, c = 1$
$$x = \dfrac{-(-7) \pm \sqrt{(-7)^2 - 4 \times 4 \times 1}}{2 \times 4} = \dfrac{7 \pm \sqrt{49 - 16}}{8} = \dfrac{7 \pm \sqrt{33}}{8}$$
So $x = \dfrac{7 + \sqrt{33}}{8} = \mathbf{1.593}$ **(3 d.p.)** and
$x = \dfrac{7 - \sqrt{33}}{8} = \mathbf{0.157}$ **(3 d.p.)**

b) $6x^2 + x = 4 \Rightarrow 6x^2 + x - 4 = 0 \Rightarrow a = 6, b = 1, c = -4$
$$x = \dfrac{-1 \pm \sqrt{1^2 - 4 \times 6 \times -4}}{2 \times 6} = \dfrac{-1 \pm \sqrt{1 + 96}}{12} = \dfrac{-1 \pm \sqrt{97}}{12}$$
So $x = \dfrac{\sqrt{97} - 1}{12} = \mathbf{0.737}$ **(3 d.p.)** and
$x = \dfrac{-\sqrt{97} - 1}{12} = \mathbf{-0.904}$ **(3 d.p.)**

c) $-2x^2 = 3x - 4 \Rightarrow 2x^2 + 3x - 4 = 0 \Rightarrow a = 2, b = 3, c = -4$
$$x = \dfrac{-3 \pm \sqrt{3^2 - 4 \times 2 \times -4}}{2 \times 2} = \dfrac{-3 \pm \sqrt{9 + 32}}{4} = \dfrac{-3 \pm \sqrt{41}}{4}$$
So $x = \dfrac{\sqrt{41} - 3}{4} = \mathbf{0.851}$ **(3 d.p.)** and
$x = \dfrac{-\sqrt{41} - 3}{4} = \mathbf{-2.351}$ **(3 d.p.)**

2 $5x^2 + 7x + 3$ has discriminant $7^2 - 4 \times 5 \times 3 = \mathbf{-11}$.
This is negative, so the quadratic has **no real roots**.

Answers

Page 19 — Completing the Square

1 a) $x^2 + 4x + 1 \Rightarrow$ Initial bracket is $(x + 2)^2 = x^2 + 4x + 4$
   The adjusting number is $1 - 4 = -3$
   $\Rightarrow x^2 + 4x + 1 = (x + 2)^2 - 3$
 b) $x^2 - 12x + 5 \Rightarrow$ Initial bracket is $(x - 6)^2 = x^2 - 12x + 36$
   The adjusting number is $5 - 36 = -31$
   $\Rightarrow x^2 + 12x + 5 = (x - 6)^2 - 31$
 c) $x^2 - 20x - 10 \Rightarrow$ Initial bracket is $(x - 10)^2 = x^2 - 20x + 100$
   The adjusting number is $-10 - 100 = -110$
   $\Rightarrow x^2 - 20x - 10 = (x - 10)^2 - 110$
 d) $x^2 + 7x - 3 \Rightarrow$ Initial bracket is $(x + \frac{7}{2})^2 = x^2 + 7x + \frac{49}{4}$
   The adjusting number is $-3 - \frac{49}{4} = -\frac{61}{4}$
   $\Rightarrow x^2 + 7x - 3 = (x + \frac{7}{2})^2 - \frac{61}{4}$

2 $x^2 + 6x - 8 \Rightarrow$ Initial bracket is $(x + 3)^2$
   The adjusting number is $-8 - 9 = -17$
   $\Rightarrow x^2 + 6x - 8 = (x + 3)^2 - 17$
   $(x + 3)^2 - 17 = 0 \Rightarrow (x + 3)^2 = 17$
   $\Rightarrow x + 3 = \pm\sqrt{17} \Rightarrow x = -3 + \sqrt{17}$ and $x = -3 - \sqrt{17}$

Page 20 — Completing the Square

1 a) $x^2 + 16x + 3 \Rightarrow$ Initial bracket is $(x + 8)^2$
   The adjusting number is $3 - 64 = -61$
   $\Rightarrow x^2 + 16x + 3 = (x + 8)^2 - 61$
 b) If $x^2 + 16x + 3 = 0$, then $(x + 8)^2 - 61 = 0$
   $\Rightarrow (x + 8)^2 = 61 \Rightarrow x + 8 = \pm\sqrt{61} \Rightarrow x = -8 + \sqrt{61}$ or
   $\qquad\qquad\qquad\qquad x = -8 - \sqrt{61}$
 c) The coefficient of x^2 is positive, so the graph
   is u-shaped and has a minimum point.
   This is when $(x + 8)^2 = 0$, so when $x = -8$. At this point,
   $y = -61$, so the minimum point is at $(-8, -61)$.
   The x-intercepts are the solutions to $x^2 + 16x + 3 = 0$
   from part b): $(-8 - \sqrt{61}, 0)$ and $(-8 + \sqrt{61}, 0)$
   The y-intercept is at $x = 0$: $y = 0^2 + (16 \times 0) + 3 = 3$,
   so y-intercept is $(0, 3)$.

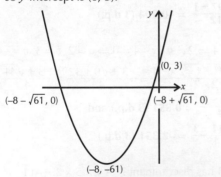

2 a) From the sketch, you can see the minimum point
   is at $(-3, 3)$ so the adjusting number is $q = 3$.
   And it tells you that $(-3 + p) = 0$, so $p = 3$.
 b) From part a), you know that $y = (x + 3)^2 + 3$.
   At the y-intercept, $x = 0$, so $y = 3^2 + 3 = 12$,
   which means the y-intercept is at $(0, 12)$.

Page 21 — Completing the Square

1 a) $2x^2 - 3x - 10 = 2(x^2 - \frac{3}{2}x) - 10$
   $\Rightarrow$ Initial bracket is $2(x - \frac{3}{4})^2 = 2x^2 - 3x + \frac{9}{8}$
   The adjusting number is $-10 - \frac{9}{8} = -\frac{89}{8}$
   $\Rightarrow 2x^2 - 3x - 10 = 2(x - \frac{3}{4})^2 - \frac{89}{8}$
 b) $-x^2 - 3x - 1 = -(x^2 + 3x) - 1$
   $\Rightarrow$ Initial bracket is $-(x + \frac{3}{2})^2 = -x^2 - 3x - \frac{9}{4}$
   The adjusting number is $-1 - \left(-\frac{9}{4}\right) = -1 + \frac{9}{4} = \frac{5}{4}$
   $\Rightarrow -x^2 - 3x - 1 = -(x + \frac{3}{2})^2 + \frac{5}{4}$
 c) $5x^2 - x - 3 = 5(x^2 - \frac{1}{5}x) - 3$
   $\Rightarrow$ Initial bracket is $5(x - \frac{1}{10})^2 = 5x^2 - x + \frac{1}{20}$
   The adjusting number is $-3 - \frac{1}{20} = -\frac{61}{20}$
   $\Rightarrow 5x^2 - x - 3 = 5(x - \frac{1}{10})^2 - \frac{61}{20}$
 d) $-\frac{1}{2}x^2 - 3x - 7 = -\frac{1}{2}(x^2 + 6x) - 7$
   $\Rightarrow$ Initial bracket is $-\frac{1}{2}(x + 3)^2 = -\frac{1}{2}x^2 - 3x - \frac{9}{2}$
   The adjusting number is $-7 - \left(-\frac{9}{2}\right) = -\frac{5}{2}$
   $\Rightarrow -\frac{1}{2}x^2 - 3x - 7 = -\frac{1}{2}(x + 3)^2 - \frac{5}{2}$

2 $x^2 + y^2 - 2x + 6y - 15 = x^2 - 2x + y^2 + 6y - 15 = 0$
   $\Rightarrow (x - 1)^2 - 1 + (y + 3)^2 - 9 - 15 = 0$
   $\Rightarrow (x - 1)^2 + (y + 3)^2 - 25 = 0$
   $\Rightarrow (x - 1)^2 + (y + 3)^2 = 25$

Section 4 — More Algebra

Page 22 — Algebraic Fractions

1 a) $\dfrac{3x^2}{7x} = \dfrac{3x}{7}$
 b) $\dfrac{8x + 16}{2x - 4} = \dfrac{8(x + 2)}{2(x - 2)} = \dfrac{4(x + 2)}{x - 2}$
 c) $\dfrac{x^2 - 25}{5(x + 5)} = \dfrac{(x + 5)(x - 5)}{5(x + 5)} = \dfrac{x - 5}{5}$
 d) $\dfrac{3x^2 + 16x - 12}{2x^2 + 13x + 6} = \dfrac{(3x - 2)(x + 6)}{(2x + 1)(x + 6)} = \dfrac{3x - 2}{2x + 1}$
 e) $\dfrac{x^3 + 2x^2 + x}{x^2 - 3x - 4} = \dfrac{x(x^2 + 2x + 1)}{(x - 4)(x + 1)} = \dfrac{x(x + 1)^2}{(x - 4)(x + 1)} = \dfrac{x(x + 1)}{x - 4}$

2 a) $\dfrac{x + 3}{x^2} \times \dfrac{x}{4} = \dfrac{x + 3}{x} \times \dfrac{1}{4} = \dfrac{x + 3}{4x}$
 b) $\dfrac{3x + 9}{4} \times \dfrac{x}{3(x + 3)} = \dfrac{3(x + 3)}{4} \times \dfrac{x}{3(x + 3)} = \dfrac{1}{4} \times \dfrac{x}{1} = \dfrac{x}{4}$
 c) $\dfrac{10x}{x^2 - 9} \div \dfrac{2x + 14}{x - 3} = \dfrac{10x}{x^2 - 9} \times \dfrac{x - 3}{2x + 14}$
   $\qquad = \dfrac{10x}{(x - 3)(x + 3)} \times \dfrac{x - 3}{2(x + 7)}$
   $\qquad = \dfrac{5x}{x + 3} \times \dfrac{1}{x + 7} = \dfrac{5x}{(x + 3)(x + 7)}$
 d) $\dfrac{x^2 + 5x + 6}{9} \div \dfrac{x^2 - 4x - 21}{6x - 42} = \dfrac{x^2 + 5x + 6}{9} \times \dfrac{6x - 42}{x^2 - 4x - 21}$
   $\qquad = \dfrac{(x + 3)(x + 2)}{9} \times \dfrac{6(x - 7)}{(x - 7)(x + 3)}$
   $\qquad = \dfrac{x + 2}{3} \times \dfrac{2}{1} = \dfrac{2(x + 2)}{3}$

Answers

Page 23 — Algebraic Fractions

1 a) $3 + \dfrac{2}{x} = \dfrac{3x}{x} + \dfrac{2}{x} = \dfrac{3x+2}{x}$

b) $\dfrac{a}{b} - \dfrac{2a}{3b} = \dfrac{3a}{3b} - \dfrac{2a}{3b} = \dfrac{a}{3b}$

c) $\dfrac{1}{x+1} - \dfrac{3}{x} = \dfrac{x}{x(x+1)} - \dfrac{3(x+1)}{x(x+1)} = \dfrac{x-3x-3}{x(x+1)} = \dfrac{-2x-3}{x(x+1)}$

d) $\dfrac{4}{x+2} + \dfrac{3}{x^2} = \dfrac{4x^2}{x^2(x+2)} + \dfrac{3(x+2)}{x^2(x+2)} = \dfrac{4x^2+3x+6}{x^2(x+2)}$

e) $\dfrac{1}{x+1} - \dfrac{1}{x(x+1)} + \dfrac{1}{x} = \dfrac{x}{x(x+1)} - \dfrac{1}{x(x+1)} + \dfrac{x+1}{x(x+1)}$

$\qquad = \dfrac{x-1+x+1}{x(x+1)} = \dfrac{2x}{x(x+1)}$

f) $\dfrac{5x}{7} + \dfrac{x+3}{2x} = \dfrac{10x^2}{14x} + \dfrac{7(x+3)}{14x} = \dfrac{10x^2+7x+21}{14x}$

g) $\dfrac{x-1}{x+8} - \dfrac{3}{x+1} = \dfrac{(x-1)(x+1)}{(x+8)(x+1)} - \dfrac{3(x+8)}{(x+8)(x+1)}$

$\qquad = \dfrac{x^2-1-3x-24}{(x+8)(x+1)} = \dfrac{x^2-3x-25}{(x+8)(x+1)}$

h) $1 + \dfrac{4}{x-4} - \dfrac{2}{x+2}$

$\qquad = \dfrac{(x-4)(x+2)}{(x-4)(x+2)} + \dfrac{4(x+2)}{(x-4)(x+2)} - \dfrac{2(x-4)}{(x-4)(x+2)}$

$\qquad = \dfrac{x^2-2x-8+4x+8-2x+8}{(x-4)(x+2)} = \dfrac{x^2+8}{(x-4)(x+2)}$

Page 25 — Inequalities

1 a) $4 - 8x \le -10x - 6 \Rightarrow 2x \le -10 \Rightarrow \mathbf{x \le -5}$

b) $-13 \le 2x - 3 < 11 \Rightarrow -10 \le 2x < 14 \Rightarrow \mathbf{-5 \le x < 7}$

c) $-11 < 1 - 3x < 7 \Rightarrow -12 < -3x < 6 \Rightarrow \mathbf{-2 < x < 4}$

2 $\quad 2x - 4 < 3x \Rightarrow x > -4$

$\quad 5 \ge 7x - 2 \Rightarrow 7x \le 7 \Rightarrow x \le 1$

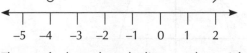

The set of values where the lines overlap satisfy both
inequalities, so $\mathbf{-4 < x \le 1}$.

3 a) $3x^2 - 8 < 4 \Rightarrow 3x^2 < 12 \Rightarrow x^2 < 4 \Rightarrow \mathbf{-2 < x < 2}$

b) $2x - x^2 + 8 > 0 \Rightarrow x^2 - 2x - 8 < 0$
The graph of $y = x^2 - 2x - 8 = (x-4)(x+2)$ is a u-shaped
curve that crosses the x-axis at $x = 4$ and $x = -2$, and the
curve is below the x-axis between these points,
so $\mathbf{-2 < x < 4}$.

c) $6x \ge 2x^2 + 4 \Rightarrow 2x^2 - 6x + 4 \le 0$
The graph of $y = 2x^2 - 6x + 4 = 2(x-1)(x-2)$ is a u-shaped
curve that crosses the x-axis at $x = 1$ and $x = 2$, and
the curve is below the x-axis between these points,
so $\mathbf{1 \le x \le 2}$.

d) $-5x^2 + 4x + 2 < 6x - x^2 \Rightarrow 4x^2 + 2x - 2 > 0$
$\qquad\qquad\qquad\qquad\qquad \Rightarrow 2x^2 + x - 1 > 0$
The graph of $y = 2x^2 + x - 1 = (2x-1)(x+1)$ is a u-shaped
curve that crosses the x-axis at $x = -1$ and
$x = \dfrac{1}{2}$. The curve is above the x-axis when x is less than
-1 and when x is more than $\dfrac{1}{2}$,
so $\mathbf{x < -1}$ or $\mathbf{x > \dfrac{1}{2}}$.

4 $\quad x^2 + 3x + 7 < 3 - 2x \Rightarrow x^2 + 5x + 4 < 0$
The graph of $y = x^2 + 5x + 4 = (x+4)(x+1)$ is a
u-shaped curve that crosses the x-axis at $x = -4$ and $x = -1$,
and the curve is below the x-axis between
these points, so $-4 < x < -1$.

$2(3 - 2x) > 14 \Rightarrow 6 - 4x > 14 \Rightarrow 4x < -8 \Rightarrow x < -2$

So the values of x which satisfy both
inequalities are $\mathbf{-4 < x < -2}$.

Page 26 — Graphical Inequalities

1 a) Write the inequality as an equation in the form '$y =$':
$y = 17 - 5x$. Draw the graph of $y = 17 - 5x$.
Decide which side of the line is correct by plugging
$(0, 0)$ into the inequality: $5x + y > 17 \Rightarrow 0 > 17$ which is
false. So the origin is on the wrong side of the line.
So shade the region above $y = 17 - 5x$.

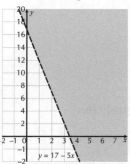

b) Write the inequality as an equation in the form '$y =$':
$y = 1 - 4x$. Draw the graph of $y = 1 - 4x$. Use the method
from part a) to find and shade the region:

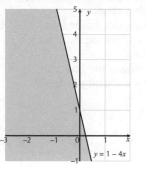

c) Write the inequality as an equation in the form '$y =$':
$y = \dfrac{1}{2}x - 4$. Draw the graph of $y = 17 - 5x$.
Use the method from part a) to find and shade the region:

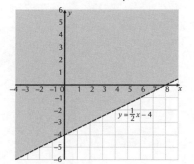

2   Write the second and third inequalities as equations in the form '$y = $': $y = 2 - x$ and $y = x - 10$.
Draw the graphs of $x = 8$, $y = 2 - x$ and $y = x - 10$.

Decide which side of each line is correct by plugging $(0, 0)$ into each inequality: $x < 8 \Rightarrow 0 < 8$ which is true.
So the origin is on the correct side of the line.
$x + y > 2 \Rightarrow 0 > 2$ which is false.
So the origin is on the wrong side of the line.
$x - 4 < y + 6 \Rightarrow -4 < 6$ which is true.
So the origin is on the correct side of the line.
So shade the region to the left of $x = 8$ and above $y = 2 - x$ and $y = x - 10$.

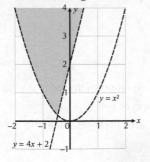

Page 27 — Graphical Inequalities

1 a) Write each inequality as an equation in the form '$y = $':
$y = x^2$ and $y = 4x + 2$
Draw the graphs of $y = x^2$ and $y = 4x + 2$.
$y = x^2$ is u-shaped and intersects the axes at the origin.
Find the region by substituting coordinates into each inequality. Use the origin for $y = 4x + 2$:
$8x + 4 < 2y \Rightarrow 4 < 0$ which is false.
So the origin is on the wrong side of the line.
The origin is on the line $y = x^2$ so use $(1, 2)$ for $y > x^2$:
$y > x^2 \Rightarrow 2 > 1$ which is true.
So $(1, 2)$ is on the correct side of the line.
So label the region above $y = x^2$ and $y = 4x + 2$.

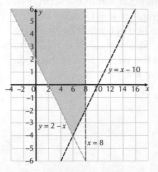

b) Write each inequality as an equation in the form '$y = $':
$y = -x^2 + 4$ and $y = 4 - x$
Draw the graphs of $y = -x^2 + 4$ and $y = 4 - x$.
$y = -x^2 + 4$ is n-shaped. To find where it crosses the x-axis, solve $-x^2 + 4 = 0$: $-x^2 + 4 = 0 \Rightarrow x^2 - 4 = 0$
$\Rightarrow (x + 2)(x - 2) = 0$ so the graph crosses the x-axis when $x = 2$ and $x = -2$. The x-coordinate of the turning point is halfway between 2 and -2, so at $x = 0$.
It also crosses the y-axis when $x = 0$, so $y = 4$.
This also gives the y-coordinate of the turning point as 4.
Find the region by substituting $(0, 0)$ into each inequality:
$-x^2 + 4 \geq y \Rightarrow 4 \geq 0$ which is true.
So the origin is on the correct side of the line.
$y + 10 < 14 - x \Rightarrow 10 < 14$ which is true.
So the origin is on the correct side of the line.
So label the region below $y = -x^2 + 4$ and $y = 4 - x$.

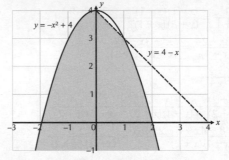

c) Write each inequality as an equation in the form '$y = $':
$y = 8 - 2x^2$ and $y = 2x + 6$
Draw the graphs of $y = 8 - 2x^2$ and $y = 2x + 6$. Use the method from part b) to find and label the region:

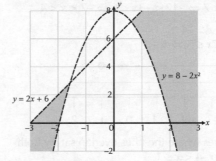

d) Write each inequality as an equation in the form '$y = $':
$y = -\dfrac{x^2}{4}$ and $y = x^2 - 4$.
Draw the graphs of $y = -\dfrac{x^2}{4}$ and $y = x^2 - 4$.
Use the method from part b) to find and label the region:

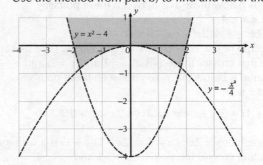

Answers

2   Write each inequality as an equation in the form '$y =$':
$y = -2x^2 + 5x + 3$ and $y = x^2 - x - 2$.
Draw the graphs of $y = -2x^2 + 5x + 3$ and $y = x^2 - x - 2$.
Use the method from 1b) to find and shade the region:

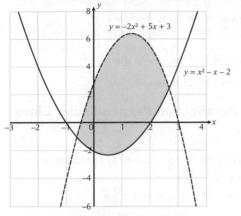

Page 28 — Simultaneous Equations

1 a) Write the equations in the correct form and label them:
(1) $4x - y = 1$ and (2) $5x - 2y = 5$
(1) × 2: $8x - 2y = 2$ (3)
(3) − (2): $3x = -3 \Rightarrow x = -1$
Sub into (1): $(4 \times -1) - y = 1 \Rightarrow -4 - y = 1 \Rightarrow y = -5$
Solution: $x = -1, y = -5$

   b) Label the equations:
(1) $3x + 8y = 25$ and (2) $12x - 10y = 16$
(1) × 4: $12x + 32y = 100$ (3)
(3) − (2): $42y = 84 \Rightarrow y = 2$
Sub into (1): $3x + (8 \times 2) = 25 \Rightarrow 3x + 16 = 25$
$\Rightarrow 3x = 9 \Rightarrow x = 3$
Solution: $x = 3, y = 2$

   c) Label the equations:
(1) $3x - 2y = 8$ and (2) $7x - 5y = 17$
(1) × 5: $15x - 10y = 40$ (3)
(2) × 2: $14x - 10y = 34$ (4)
(3) − (4): $x = 6$
Sub into (1): $(3 \times 6) - 2y = 8 \Rightarrow 18 - 2y = 8$
$\Rightarrow 2y = 10 \Rightarrow y = 5$
Solution: $x = 6, y = 5$

   d) Write the equations in the correct form and label them:
(1) $\frac{1}{2}x + 2y = 12$ and (2) $4x - y = 11$
(1) × 8: $4x + 16y = 96$ (3)
(3) − (2): $17y = 85 \Rightarrow y = 5$
Sub into (1): $\frac{1}{2}x + (2 \times 5) = 12 \Rightarrow \frac{1}{2}x + 10 = 12$
$\Rightarrow \frac{1}{2}x = 2 \Rightarrow x = 4$
Solution: $x = 4, y = 5$

Page 29 — Simultaneous Equations

1 a) Label the equations:
(1) $y = 4x + 4$ and (2) $y = x^2 + 3x - 8$
Substitute (1) into (2): $4x + 4 = x^2 + 3x - 8$
$x^2 - x - 12 = 0$
$(x - 4)(x + 3) = 0$
$x = 4$ or $x = -3$
Substitute $x = 4$ into (1): $y = (4 \times 4) + 4 = 20$
Substitute $x = -3$ into (1): $y = (4 \times -3) + 4 = -8$
Solutions: $x = 4, y = 20$ and $x = -3, y = -8$

   b) Rearrange the linear equation to get y
on its own and label the equations:
(1) $y = 3x - 3$ and (2) $y^2 - x^2 = 0$
Substitute (1) into (2): $(3x - 3)^2 - x^2 = 0$
$8x^2 - 18x + 9 = 0$
Use the quadratic formula with $a = 8$, $b = -18$ and $c = 9$:
$x = \dfrac{-(-18) \pm \sqrt{(-18)^2 - 4 \times 8 \times 9}}{2 \times 8} = \dfrac{18 \pm \sqrt{36}}{16} = \dfrac{18 \pm 6}{16}$
$x = 1.5$ or $x = 0.75$
Substitute $x = 1.5$ into (1): $y = (3 \times 1.5) - 3 = 1.5$
Substitute $x = 0.75$ into (1): $y = (3 \times 0.75) - 3 = -0.75$
Solutions: $x = 1.5, y = 1.5$ and $x = 0.75, y = -0.75$

   c) Rearrange the linear equation to get y
on its own and label the equations:
(1) $y = -2x - 7$ and (2) $x^2 + xy - 10 = 0$
Substitute (1) into (2): $x^2 + x(-2x - 7) - 10 = 0$
$-x^2 - 7x - 10 = 0$
$x^2 + 7x + 10 = 0$
$(x + 5)(x + 2) = 0$
$x = -5$ or $x = -2$
Substitute $x = -5$ into (1): $y = (-2 \times -5) - 7 = 3$
Substitute $x = -2$ into (1): $y = (-2 \times -2) - 7 = -3$
Solutions: $x = -5, y = 3$ and $x = -2, y = -3$

Page 31 — Proof

1   $(n + 1)^2 - (n - 1)^2 = n^2 + 2n + 1 - (n^2 - 2n + 1)$
$= 4n$
$= 2(2n)$
$(n + 1)^2 - (n - 1)^2$ can be written as $2x$, where $x = 2n$ is an integer, so it is even.

2   Call the two odd numbers $2a + 1$ and $2b + 1$.
$(2a + 1)(2b + 1) = 4ab + 2a + 2b + 1$
$= 2(ab + a + b) + 1$
The product of two odd numbers can be written as $2x + 1$, where x is an integer and $x = ab + a + b$, so is odd.

3   Call the two rational numbers x and y. Rational numbers can be written as fractions with integers on top and bottom, so write $x = \frac{a}{b}$ and $y = \frac{c}{d}$ where a, b, c and d are integers, and b and d are not 0 (because you can't divide by 0).
$\frac{a}{b} - \frac{c}{d} = \frac{ad}{bd} - \frac{bc}{bd} = \frac{ad - bc}{bd}$. ad and bc are the product of integers, so are also integers. So $ad - bc$ is an integer.
bd is also an integer as it is the product of two integers.
It is not 0 because neither b nor d are 0.
The result of subtracting two rational numbers can be written as a fraction with integers on the top and bottom, so it is rational.

4   Try $n = 3$:
$2^3 + 1 = 8 + 1 = 9$. 9 is not prime, so Phyllis is **wrong**.

5   If $x = 1$ and $y = -2$, then $\frac{x}{y} = -\frac{1}{2}$ and $\frac{x^2}{y^2} = \frac{1}{4}$.
This means $\frac{x}{y} < \frac{x^2}{y^2}$ so the statement is false.

Answers

Page 33 — Functions

1 $g(2) = \dfrac{(11 \times 2) + 13}{21 - (7 \times 2)} = \dfrac{35}{7} = \mathbf{5}$

2 $f(3) = (5 \times 3^2) + (12 \times 3) - 7 = 45 + 36 - 7 = \mathbf{74}$

3 a) $f(2) = (6 \times 2) + 4 = 12 + 4 = \mathbf{16}$

   b) $g(4) = \dfrac{28}{4} = \mathbf{7}$

   c) $fg(x) = f\left(\dfrac{28}{x}\right) = \left(6 \times \dfrac{28}{x}\right) + 4 = \dfrac{\mathbf{168}}{\mathbf{x}} + \mathbf{4}$

   d) $gf(x) = g(6x + 4) = \dfrac{28}{6x + 4} = \dfrac{\mathbf{14}}{\mathbf{3x + 2}}$

4 a) $g(0) = (6 \times 0) - 6 = -6$
   $\Rightarrow fg(0) = f(-6) = (4 \times (-6)^2) + 3 = \mathbf{147}$
   b) $f(0) = (4 \times 0^2) + 3 = 3 \Rightarrow gf(0) = g(3) = (6 \times 3) - 6 = \mathbf{12}$
   c) $f(0) = 3 \Rightarrow f^2(0) = ff(0) = f(3) = (4 \times 3^2) + 3 = \mathbf{39}$
   d) $g(0) = -6 \Rightarrow g^2(0) = gg(0) = g(-6) = (6 \times -6) - 6 = \mathbf{-42}$

5 a) $x = 9 - 11y \Rightarrow y = \dfrac{9 - x}{11}$ so $f^{-1}(x) = \dfrac{\mathbf{9 - x}}{\mathbf{11}}$

   b) $x = \dfrac{y - 4}{2} \Rightarrow y = 2x + 4$ so $g^{-1}(x) = \mathbf{2x + 4}$

Section 5 — Graphs

Page 34 — Straight Lines

1 Rearrange the equation of the line into
   $y = mx + c$ form to find the gradient:
   $2x + y = -2 \Rightarrow y = -2x - 2$, so the gradient is **–2**.
   When $y = 0$, $2x = -2 \Rightarrow x = -1$, so x-intercept is **(–1, 0)**.
   When $x = 0$, $y = -2$, so y-intercept is **(0, –2)**.

2 $m = 3$, $c = 5$, so $\mathbf{y = 3x + 5}$.

3 $(x_1, y_1) = (3, 4)$, $(x_2, y_2) = (-2, 1)$.
   So the gradient is $\dfrac{1 - 4}{-2 - 3} = \dfrac{-3}{-5} = \dfrac{\mathbf{3}}{\mathbf{5}}$

Page 35 — Straight Lines

1 $(x_1, y_1) = (1, 2)$, $(x_2, y_2) = (5, 9)$.
   The gradient is $\dfrac{9 - 2}{5 - 1} = \dfrac{7}{4} \Rightarrow y = \dfrac{7}{4}x + c$
   Plugging in the point (1, 2) gives:
   $2 = \dfrac{7}{4} + c \Rightarrow c = 2 - \dfrac{7}{4} = \dfrac{1}{4}$, so $\mathbf{y = \dfrac{7}{4}x + \dfrac{1}{4}}$.

2 a) $(x_1, y_1) = (-3, 4)$, $(x_2, y_2) = (7, -2)$.
   The gradient is $\dfrac{-2 - 4}{7 - (-3)} = \dfrac{-6}{10} = -\dfrac{3}{5} \Rightarrow y = -\dfrac{3}{5}x + c$
   Plugging in the point (–3, 4) gives:
   $4 = -\dfrac{3}{5} \times -3 + c \Rightarrow 4 = \dfrac{9}{5} + c \Rightarrow c = 4 - \dfrac{9}{5} = \dfrac{11}{5}$
   So $y = -\dfrac{3}{5}x + \dfrac{11}{5} \Rightarrow 5y = -3x + 11$
   $\Rightarrow \mathbf{3x + 5y - 11 = 0}$

   b) $\sqrt{(7 - (-3))^2 + (-2 - 4)^2} = \sqrt{10^2 + (-6)^2} = \sqrt{136}$
   $= \mathbf{11.66}$ **to 2 d.p.**

3 a) At $t = 2$, $d = 10$, so $(t_1, d_1) = (2, 10)$.
   At $t = 3.5$, $d = 17.5$, so $(t_2, d_2) = (3.5, 17.5)$.
   Gradient of line through these two points
   $= \dfrac{d_2 - d_1}{t_2 - t_1} = \dfrac{17.5 - 10}{3.5 - 2} = \dfrac{7.5}{1.5} = 5$
   So the equation so far is $d = 5t + c$.
   Substitute (2, 10) into the equation and solve for c:
   $d = 5t + c \Rightarrow 10 = 5 \times 2 + c \Rightarrow c = 10 - 10 = 0$.
   So the equation of the line is $\mathbf{d = 5t}$.

   b) $12.5 = 5t \Rightarrow t = \mathbf{2.5}$ **hours**

Page 36 — Parallel and Perpendicular Lines

1 $5y + 3x - 7 = 0 \Rightarrow y = -\dfrac{3}{5}x + \dfrac{7}{5}$
   Line F's gradient is $-\dfrac{3}{5}$, so line G's gradient is $-\dfrac{3}{5}$
   Line G goes through the origin, so $c = 0$.
   The equation of line G is $\mathbf{y = -\dfrac{3}{5}x}$.

2 Line P is given by $y = 4x + c$. Plug in point (–1, –2):
   $-2 = -4 + c \Rightarrow c = 2$, so line P's equation is $y = 4x + 2$.
   Rearrange to get: $\mathbf{4x - y + 2 = 0}$.
   Line Q has gradient $-1 \div 4 = -\dfrac{1}{4}$, so line Q's equation is $y = -\dfrac{1}{4}x + c$. Plugging in the point (–1, –2) gives:
   $-2 = \dfrac{1}{4} + c \Rightarrow c = -\dfrac{9}{4}$. So line Q's equation is
   $y = -\dfrac{1}{4}x - \dfrac{9}{4}$. Rearrange to get: $\mathbf{x + 4y + 9 = 0}$.

3 $(x_1, y_1) = (4, 3)$, $(x_2, y_2) = (9, 5)$
   l_1's gradient is $\dfrac{5 - 3}{9 - 4} = \dfrac{2}{5}$,
   so l_2's gradient is $-1 \div \dfrac{2}{5} = -1 \times \dfrac{5}{2} = -\dfrac{5}{2}$
   l_2 is given by $y = -\dfrac{5}{2}x + c$. Plug in point Q (9, 5):
   $5 = -\dfrac{45}{2} + c \Rightarrow c = 5 + \dfrac{45}{2} = \dfrac{55}{2}$,
   so l_2 is given by $y = -\dfrac{5}{2}x + \dfrac{55}{2}$.
   Rearrange to get: $\mathbf{5x + 2y - 55 = 0}$.

Page 37 — Quadratic Graphs

1 a) The x^2 coefficient is positive, so this is a u-shaped quadratic with a minimum point.
   $x = 0 \Rightarrow y = 0^2 - 3 \times 0 = 0 \Rightarrow$ y-intercept is at (0, 0).
   Solve by factorising to find the x-intercepts:
   $x^2 - 3x = x(x - 3) = 0 \Rightarrow$ x-intercepts at (0, 0) and (3, 0).
   The minimum point is half way between the x-intercepts,
   so is at $x = \dfrac{3 + 0}{2} = \dfrac{3}{2}$.
   $x = \dfrac{3}{2} \Rightarrow y = \left(\dfrac{3}{2}\right)^2 - 3 \times \dfrac{3}{2} = \dfrac{9}{4} - \dfrac{9}{2} = -\dfrac{9}{4}$,
   so the minimum point is at $\left(\dfrac{3}{2}, -\dfrac{9}{4}\right)$ [or (1.5, –2.25)].

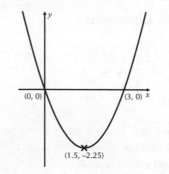

Answers

b) Use the method from part a) to sketch and label the graph.

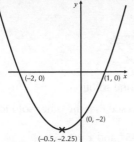

c) Use the method from part a) to sketch and label the graph. The coefficient of x^2 is negative, so the graph is n-shaped. Also, the quadratic has a double root at $x = 3$, so this is the vertex and the graph just touches the x-axis here.

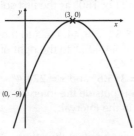

d) Use the method from part a) to sketch and label the graph.

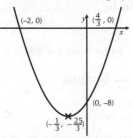

Page 38 — Harder Graphs

1 a) $y = (x + 1)(x - 1)(x + 2)$, so the x-intercepts are at $x = -1, 1$ and -2. $x = 0 \Rightarrow y = -2$, so the y-intercept is at -2.

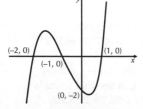

b) Use the method from part a) to sketch and label the graph.

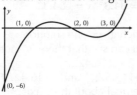

c) Factorise the cubic: $-x^3 - 5x^2 + 6x = -x(x^2 + 5x - 6) = -x(x + 6)(x - 1)$. Now use the method from part a) to sketch and label the graph. This is a negative cubic, so goes from top left to bottom right.

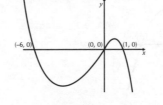

Page 39 — Harder Graphs

1 a) $a = 1, b = 3, r^2 = 16$. So the circle has centre **(1, 3)** and radius **4**.

b) $a = 0, b = -2, r^2 = 50$ So the circle has centre **(0, -2)** and radius $\sqrt{50} = \sqrt{25}\sqrt{2} = \mathbf{5\sqrt{2}}$

c) $a = -\frac{1}{2}, b = \frac{1}{2}$ and $r^2 = 2$. So the circle has centre $(-\frac{1}{2}, \frac{1}{2})$ and radius $\sqrt{2}$.

2 The centre of this circle is at (1, 2). The radius of the circle goes from (1, 2) to (3, 4), so it has gradient $\frac{4-2}{3-1} = \frac{2}{2} = 1$. This means that the tangent at (3, 4) has gradient -1. So the equation of the tangent so far is $y = -x + c$. Plug in point (3, 4) to get $4 = -3 + c \Rightarrow c = 7$, so the equation of the tangent is $y = -x + 7$.

Page 40 — Graph Transformations

1 a) i)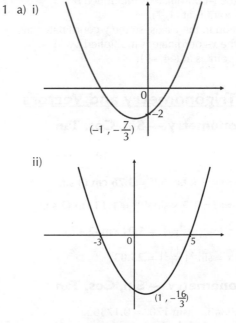
ii)

b) f(x) + 3 is described by the column vector $\begin{pmatrix} 0 \\ 3 \end{pmatrix}$.
f(x − 2) is described by the column vector $\begin{pmatrix} 2 \\ 0 \end{pmatrix}$.

Answers

Page 41 — Graph Transformations

1   The graph of $y = -f(x)$ looks like this:

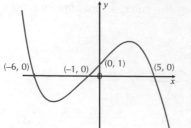

The graph of $y = f(-x)$ looks like this:

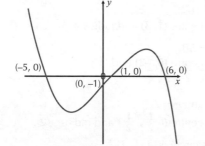

2 a) $g(x) = g(-x)$ means that the graph is symmetrical about the y-axis. $x^2 - 4$ **could** be $g(x)$, because $y = x^2 - 4$ is symmetrical about the y-axis.

  b) This **cannot** be $g(x)$ — it is not symmetrical about the y-axis.

  c) This **could** be $g(x)$, because $-x^2$ is symmetrical about the y-axis.

3 a) This is a reflection in the x-axis, so the x-coordinate stays the same, but the y-coordinate is multiplied by –1. So the turning point is at **(4, 7)**.

  b) This is a reflection in the y-axis, so the y-coordinate stays the same, but the x-coordinate is multiplied by –1. So the turning point is at **(–4, –7)**.

Section 6 — Trigonometry and Vectors

Page 42 — Trigonometry — Sin, Cos, Tan

1 a) $\cos 70° = \frac{l}{12} \Rightarrow l = 12 \times \cos 70° =$ **4.10 cm (3 s.f.)**

  b) $\tan 62° = \frac{l}{2} \Rightarrow l = 2 \times \tan 62° =$ **3.76 cm (3 s.f.)**

  c) $\sin 49° = \frac{l}{1.5} \Rightarrow l = 1.5 \times \sin 49° =$ **1.13 cm (3 s.f.)**

  d) $\cos 37° = \frac{4}{l} \Rightarrow l = \frac{4}{\cos 37°} =$ **5.01 cm (3 s.f.)**

2 $\sin x = \frac{11}{26} \Rightarrow x = \sin^{-1}\left(\frac{11}{26}\right) =$ **25.03°**

Page 43 — Trigonometry — Sin, Cos, Tan

1 $\cos 170° = -0.9848...$, $\sin 170° = 0.1736...$
So **Q = (–0.98, 0.17) (2 d.p.)**

Page 44 — Trigonometry — Graphs

1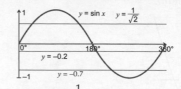

 a) $\sin 45° = \frac{1}{\sqrt{2}}$, so 45° is one solution.
Using the sketch, you can see that the other solution is at:
$180° - 45° = 135°$
So the solutions are $x =$ **45°** and $x =$ **135°**.

 b) $\sin^{-1}(-0.7) = -44.42...°$. This is outside the interval so add 360° to get the first solution in the interval:
$-44.42...° + 360 = 315.57...°$
Using the sketch above, you can see that the other solution is the same distance to the right of 180° as the first solution is to the left of 360°.
The first solution is $360 - 315.57...° = 44.42...°$ to the left of 360, so the other solution is 44.42...° to the right of 180°: $180° + 44.42...° = 224.42...°$
So solutions to 1 d.p. are $x =$ **315.6°** and $x =$ **224.4°**.

 c) $\sin^{-1}(-0.2) = -11.53...°$. This is outside the interval, so add 360° to get the first solution in the interval:
$-11.53...° + 360 = 348.46...°$.
Using the sketch above, you can see that the other solution is the same distance to the right of 180° as the first solution is to the left of 360°.
The first solution is $360 - 348.46...° = 11.53...°$ to the left of 360°, so the other solution is 11.53...° to the right of 180°: $180° + 11.53...° = 191.53...°$
So solutions to 1 d.p. are $x =$ **348.5°** and $x =$ **191.5°**.

Page 45 — Trigonometry — Graphs

1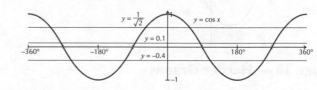

 a) $\cos 45° = \frac{1}{\sqrt{2}}$, so 45° is one solution.
Using the sketch, you can see that there is another solution at: $360° - 45° = 315°$
There are also solutions at –45° and –315° because the graph is symmetrical about the y-axis.
So the solutions are $x =$ **45°**, $x =$ **–45°**, $x =$ **315°** and $x =$ **–315°**.

 b) $\cos^{-1}(0.1) = 84.26...°$, which gives one solution.
Using the sketch above, you can see that there is another solution at $360° - 84.26...° = 275.73...°$.
There also solutions at –84.26...° and –275.73...° because the graph is symmetrical about the y-axis.
So solutions to 1 d.p. are $x =$ **84.3°**, $x =$ **–84.3°**, $x =$ **275.7°** and $x =$ **–275.7°**.

 c) $\cos^{-1}(-0.4) = 113.57...°$, which gives one solution.
Using the sketch above, you can see that there is another solution at: $360° - 113.57...° = 246.42...°$
There are also solutions at –113.57...° and –246.42...° because the graph is symmetrical about the y-axis.
So solutions to 1 d.p. are $x =$ **113.6°**, $x =$ **–113.6°**, $x =$ **246.4°** and $x =$ **–246.4°**.

Answers

Page 46 — Trigonometry — Graphs

1

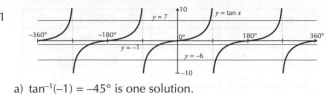

a) $\tan^{-1}(-1) = -45°$ is one solution.
Using the sketch, you can see that there are other solutions
at: $-180° - 45° = -225°$
$180° - 45° = 135°$
$360° - 45° = 315°$.
So the solutions are $x = -45°$, $x = -225°$, $x = 135°$
and $x = 315°$.

b) $\tan^{-1}(7) = 81.86...°$, which gives one solution.
Using the sketch above, you can see that there are other
solutions at: $-180° + 81.86...° = -98.13...°$
$-360° + 81.86...° = -278.13...°$
$180° + 81.86...° = 261.86...°$.
So solutions to 1 d.p. are $x = 81.9°$, $x = -98.1°$,
$x = -278.1°$ and $x = 261.9°$.

c) $\tan^{-1}(-6) = -80.53...°$, which gives one solution.
Using the sketch above, you can see that there are other
solutions at: $180° - 80.53...° = 99.46...°$
$360° - 80.53...° = 279.46...°$
$-180° - 80.53...° = -260.53...°$
So solutions to 1 d.p. are $x = -80.5°$, $x = 99.5°$,
$x = 279.5°$ and $x = -260.5°$.

Page 47 — The Sine and Cosine Rules

1 a) Using the sine rule $\dfrac{x}{\sin 42°} = \dfrac{12}{\sin 33°}$
$\Rightarrow x = \dfrac{12 \times \sin 42°}{\sin 33°} = \textbf{14.7 cm (1 d.p.)}$

b) The angle opposite x is $180° - 23° - 101° = 56°$.
Using the sine rule gives $\dfrac{x}{\sin 56°} = \dfrac{3.5}{\sin 101°}$
$\Rightarrow x = \dfrac{3.5 \times \sin 56°}{\sin 101°} = \textbf{3.0 m (1 d.p.)}$

c) Use the sine rule to find the angle that is opposite
the 31 cm side (call it 'A'): $\dfrac{31}{\sin A} = \dfrac{46}{\sin 107°}$
$\Rightarrow \sin A = \dfrac{31 \times \sin 107°}{46} \Rightarrow A = \sin^{-1}\left(\dfrac{31 \times \sin 107°}{46}\right)$
$A = 40.12...°$

So the angle opposite x is
$180° - 107° - 40.12...° = 32.87...°$
Use the sine rule again: $\dfrac{x}{\sin 32.87...°} = \dfrac{46}{\sin 107°}$
$\Rightarrow x = \dfrac{46 \times \sin 32.87...°}{\sin 107°} = \textbf{26.1 cm (1 d.p.)}$

Page 48 — The Sine and Cosine Rules

1 a) The cosine rule gives
$x^2 = 15^2 + 16^2 - 2 \times 15 \times 16 \times \cos 19°$
$\Rightarrow x = \sqrt{27.15....} = \textbf{5.2 mm (1 d.p.)}$
The area of the triangle is:
$\dfrac{1}{2} \times 15 \times 16 \times \sin 19° = \textbf{39.1 mm}^2 \textbf{ (1 d.p.)}$

b) Rearrange the cosine rule into the correct form
to find an angle: $\cos A = \dfrac{b^2 + c^2 - a^2}{2bc}$
This gives $\cos x = \dfrac{7^2 + 3^2 - 5^2}{2 \times 7 \times 3} = 0.7857...$
$\Rightarrow x = \cos^{-1}(0.7857...) = 38.21...° = \textbf{38.2° (1 d.p.)}$
The area of the triangle is:
$\dfrac{1}{2} \times 3 \times 7 \times \sin 38.21...° = \textbf{6.5 cm}^2 \textbf{ (1 d.p.)}$

c) The cosine rule gives:
$\cos x = \dfrac{11^2 + 12^2 - 13^2}{2 \times 11 \times 12} = 0.363...$
$\Rightarrow x = \cos^{-1}(0.363...) = 68.67...° = \textbf{68.7° (1 d.p.)}$
The area of the triangle is:
$\dfrac{1}{2} \times 12 \times 11 \times \sin 68.67...° = \textbf{61.5 cm}^2 \textbf{ (1 d.p.)}$

Page 49 — Vectors

1 $\overrightarrow{AC} = \overrightarrow{AB} + \overrightarrow{BC} = 2s + t + 2t - \dfrac{1}{2}s = \dfrac{3}{2}s + 3t$.
So $2\overrightarrow{AC} = 3s + 6t = 3(s + 2t)$ as required.

2 $\overrightarrow{TW} = \overrightarrow{TU} + \overrightarrow{UV} + \overrightarrow{VW} = a + b - (-2a)$
$\phantom{\overrightarrow{TW} = \overrightarrow{TU} + \overrightarrow{UV} + \overrightarrow{VW} } = a + 2a + b = 3a + b$

Page 50 — Vectors

1 $v = 4a + 6b$, $u = 6a + 9b$.
$v = 2(2a + 3b)$, $u = 3(2a + 3b)$ so $v = \dfrac{2}{3}u$.
**v and u are scalar multiples of each other,
which means they are parallel.**

2 To show the W, X and Y are collinear, you need to show
$\overrightarrow{YX}$ is parallel to $\overrightarrow{XW}$.
$\overrightarrow{YX} = \overrightarrow{YO} + \overrightarrow{OX} = -2v + u - v = u - 3v$
$\overrightarrow{XW} = \overrightarrow{XO} + \overrightarrow{OW} = -(u - v) + 2u - 4v = u - 3v$
This shows that $\overrightarrow{YX}$ is parallel to $\overrightarrow{XW}$ and
W, X and Y are collinear.

3 a) $\sqrt{3^2 + 2^2} = \sqrt{13}$
b) $\sqrt{(-1)^2 + (-1)^2} = \sqrt{2}$

Section 7 — Statistics and Probability

Page 51 — Sampling

1 E.g. This sample may be biased because it hasn't been
chosen at random from residents of the town. / The
sampling method excludes residents who don't know
about the meeting, or who are unable to attend, e.g.
because they are at work. Attendees are therefore more
likely to be older or unemployed. / It's possible that people
who live outside the town may also attend the meeting,
which would lead to sampling from the wrong population.
/ It's likely to be biased towards people with strong views,
as they are more likely to choose to attend the meeting. /
If only a small number of people attend the meeting, the
sample will be too small to represent the population.

Page 52 — Data Basics

1

Length (mm)	Frequency	Lower Class B'dary (mm)	Upper Class B'dary (mm)	Class Width (mm)	Class Mid-point (mm)
80-110	15	**75**	**115**	**40**	**95**
120-200	16	**115**	**205**	**90**	**160**
210-320	19	**205**	**325**	**120**	**265**

Answers

Page 53 — Histograms

1 a)

Height (h, cm)	Frequency	Frequency density
$150 < h \leq 170$	40	$40 \div 20 = 2$
$170 < h \leq 190$	32	$32 \div 20 = 1.6$
$190 < h \leq 210$	26	$26 \div 20 = 1.3$
$210 < h \leq 240$	9	$9 \div 30 = 0.3$

Now draw the histogram:

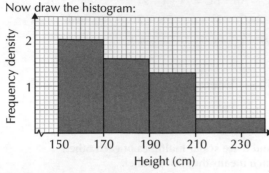

b) Between 155 cm and 170 cm:
$2 \times (170 - 155) = 2 \times 15 = 30$
All of the $170 < h \leq 190$ class is included,
which is another 32 people.
Between 190 cm and 200 cm:
$1.3 \times (200 - 190) = 1.3 \times 10 = 13$.
So the estimate is $30 + 32 + 13 = $ **75 people**.

Page 54 — Averages

1 Put the data into ascending order:
1, 4, 5, 5, 6, 7, 8
The **mode** is **5**.
There are 7 data values, so the middle value is the
fourth one, so the **median** is **5**.
Calculate the **mean**:
$(1 + 4 + 5 + 5 + 6 + 7 + 8) \div 7 = $ **5.14 (2 d.p.)**

2 Plug the values into the formula:
$\overline{x_1} = 12, n_1 = 5, \overline{x_2} = 15, n_2 = 6$
$\overline{x} = \dfrac{5 \times 12 + 6 \times 15}{5 + 6} = \dfrac{150}{11} = $ **13.64 (2 d.p.)**

Page 55 — Averages

1 a) The total of the frequencies is 30, so the median is
halfway between the 15th and 16th values.
This means the median is in the **50-90 g class**.

b)

Mass (g)	Frequency f	mid-point x	fx
0-40	6	$(45 + 0) \div 2 = 22.5$	135
50-90	11	$(45 + 95) \div 2 = 70$	770
100-140	8	$(95 + 145) \div 2 = 120$	960
150-190	5	$(145 + 195) \div 2 = 170$	850
Total	30	–	2715

So $\overline{x} = 2715 \div 30 = $ **90.5 g**

Page 56 — Cumulative Frequency

1 a)

Length (l, mm)	Frequency	Cumulative Frequency
$300 \leq l < 350$	6	6
$350 \leq l < 410$	10	$6 + 10 = 16$
$410 \leq l < 470$	24	$16 + 24 = 40$
$470 \leq l < 520$	14	$40 + 14 = 54$
$520 \leq l < 600$	6	$54 + 6 = 60$

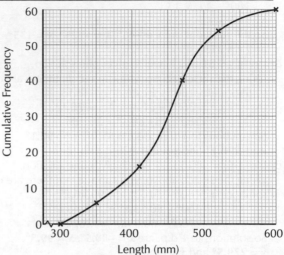

b) The median is half way up the side of the graph, so read
along from 30 to get approximately **450 mm**.
The upper quartile is 75% of the way up the side of
the graph, so read along from 45 to get approximately
482 mm.
The lower quartile is 25% of the way up the side of
the graph, so read along from 15 to get approximately
405 mm.
So the interquartile range is approximately
$482 - 405 = $ **77 mm**.

Page 57 — Probability

1 a) There are 2 outcomes where the event 'roll 1 or 2'
happens, and 6 outcomes in total.
So P(roll 1 or 2) $= \dfrac{2}{6} = \dfrac{1}{3}$.

b) There are 4 outcomes matching 'roll higher than 2',
and 6 outcomes in total.
So P(roll higher than 2) $= \dfrac{4}{6} = \dfrac{2}{3}$.
(Or you could use the fact that probabilities add up to 1.
The result of the dice roll will either be a number higher
than 2, or it will be equal to 1 or 2,
so P(roll higher than 2) $= 1 - $ P(roll 1 or 2)
$= 1 - \dfrac{1}{3} = \dfrac{2}{3}$.)

2 There are 4 outcomes where the event 'ball is yellow and
labelled A' happens, and 30 outcomes in total.
So P(ball is yellow and labelled A) $= \dfrac{4}{30} = \dfrac{2}{15}$.

Answers

Page 58 — Probability

1   You need two circles — one to show the people who bought fiction books and one to show the people who bought travel books. 2 people bought both types, so 2 goes in the overlap. 22 – 2 = 20 people bought only fiction books and 7 – 2 = 5 people bought only travel books. 40 – 20 – 2 – 5 = 13 people bought neither type of book. So the Venn diagram looks like this:

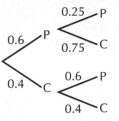

2   The event 'likes biology or chemistry or both' is shown by the area that's in B or in C or in both circles.
Add up the numbers to get:
P(likes biology or chemistry or both)
= 0.25 + 0.05 + 0.25 + 0.1 + 0.1 + 0.15 = **0.9**

Page 59 — Laws of Probability

1   If M = 'Martha plays' and S = 'Sahil plays',
then P(M or S) = P(M) + P(S) – P(M and S).
The events M and S are independent,
so P(M and S) = P(M) × P(S) = 0.3 × 0.4 = 0.12.
So P(M or S) = 0.3 + 0.4 – 0.12 = **0.58**
(Or you could do 1 – P(neither play) to get:
1 – (0.7 × 0.6) = 1 – 0.42 = 0.58)

2   You need to show that P(A) × P(B) = P(A and B).
P(A) × P(B) = 0.6 × 0.5 = 0.3 = P(A and B), so the events are independent.

Page 60 — Tree Diagrams

1   Draw a set of branches for the first Friday. At the end of each branch, draw a set for the second Friday. Claire has curry this week, so you get the following probabilities:

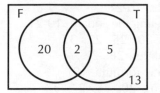

Now you can multiply along the branches P and P to find P(pizza and pizza) = 0.6 × 0.25 = **0.15**.

Index